Paul Caceres

Biota characterization and environmental management plan for the Chico River

Paul Caceres

Biota characterization and environmental management plan for the Chico River

Exploring the underwater world: characterisation of the Chichico River in Tarki

ScienciaScripts

Imprint

Cover image: www.ingimage.com

This book is a translation from the original published under ISBN 978-613-9-40559-6.

Publisher:
Sciencia Scripts
is a trademark of
Dodo Books Indian Ocean Ltd. and OmniScriptum S.R.L publishing group

120 High Road, East Finchley, London, N2 9ED, United Kingdom
Str. Armeneasca 28/1, office 1, Chisinau MD-2012, Republic of Moldova, Europe
Printed at: see last page
ISBN: 978-620-8-09361-7

Contents

THANK YOU

I express my sincere thanks to all those who made the completion of this investigation possible:

In a special way to my mother who, with her undeniable sacrifice and love, made this life project a reality, as well as to my grandparents and siblings who, with their spiritual and human support, supported me at all times.

To the National University of Loja, to the Agricultural and Livestock Area and Renewable Natural Resources, through the career of Engineering in Management and Conservation of the Environment, where I obtained the technical-scientific knowledge that has contributed to the completion of my university studies.

To the members of the thesis examining board for their valuable suggestions for this research work.

Fausto Ramiro Garcia Vasco, Mg.Sc, for his valuable suggestions in the development of the field phase, data analysis and in the direction and revision of this work.

DEDICATION

I dedicate this work with great affection to my parents who, with their example of self-improvement, instilled in me the love and passion for study, as well as giving me all the support I needed to complete my professional career and achieve my goals and objectives.

SUMMARY
In the city of Puyo, Tarqui parish, there is a micro basin of the Chichico River, where agriculture, livestock and human settlements are developed on the riverbanks, which leads to a series of impacts on the environment, such as alteration of the quality of the water resource, loss of aquatic biodiversity and illnesses in people who use this resource for domestic use; This problem is proposed to be controlled through an Environmental Management Plan, based on the information obtained from the sampling and identification of macroinvertebrates and the analysis of the physical and chemical parameters of the water of the Chichico River; The methods used for the sampling of aquatic macroinvertebrates were the surber net, the wading method for the collection of water samples for laboratory analysis, and the EPT index (Ephemeroptera, Plecoptera and Trichoptera) for the evaluation of biological indicators; The result was a total of 261 specimens, of which 145 correspond to the EPT index, which was compared with the chemical parameters of the water that are outside the norm and are related to macroinvertebrates such as the BOD with 16 mg/l and the DO with an average of 5.67 mg/l; For which an Environmental Management Plan is proposed, which contains 6 programmes to control and mitigate the existing impacts on the river.
Keywords: Macroinvertebrates, Inclice EFA, Environmental Management Plan, Programmes.

INTRODUCTION

In the micro basin of the Rto Chichico, an important water resource in the parish of Tarqui, different economic activities are developed here such as agriculture, tourism, livestock, human settlements on the banks of the river, product of the same anthropic activities, incorporating waste water, oils, grease, solid waste, etc. into the river.This causes the loss of the quality of the Indian resource and its importance as a natural water resource of the area; with the aim of proposing a solution to this problem, the characterisation of the aquatic biota was carried out to determine the quality of the Chichico River and to propose an Environmental Management Plan in the Tarqui parish".

The methodology applied to develop this research in the sampling and analysis of aquatic macroinvertebrates, was through the use of the Surber net of 30 x 30 cm surface area and 0.5 mm mesh opening and for the identification of the families of macroinvertebrates, for the evaluation of water quality, the EPT index was used to determine the presence or absence of the orders Ephemeroptera, Plecoptera and Trichoptera.

For the sampling and analysis of the physical and chemical parameters, the Ecuadorian Technical Standard NTE INEN 2 169:98 was used, which outlines the procedures for water sampling. The parameters were determined according to the Regulations of the Unified Text of Secondary Environmental Legislation (TULAS), Book VI, Table N°01, on the use of water for human consumption and domestic use.

The following results were obtained: 261 specimens were identified in the 4 sampling points, divided into 5 orders with 20 families, of which 116 belong to different orders of macroinvertebrates and 145 belong to the orders Ephemeroptera, Plecoptera and Trichoptera.

According to the Unified Text of Secondary Environmental Legislation (TULAS), both in point 03 and 04 of the BOD sampling is 16 mg/l, being the maximum permissible limits of 2 mg/l, therefore it does not comply with the established in the legislation, therefore being the BOD an indispensable parameter for the quality of the water of rivers, lakes, lagoons or effluents, it is not apt for the survival and development of the macro invertebrates, as well as for the human and domestic consumption of the surrounding population. This information made it possible to propose an Environmental Management Plan, an instrument that will serve to control and conserve the natural resources of the sub-basin of the Rto Chichico, in the parish of Tarqui.

In order to develop the research topic, the following objectives were set out:

General objective:

Characterise the aquatic biotics to determine the quality of the Chichico River and design an environmental management plan in the parish of Tarqui.
Specific objectives:

- To determine the procedure for the sampling and identification of macroinvertebrates at four points of the Rto Chichico.
- To evaluate the physical and chemical quality of the water of the Rto Chichico.
- Propose a Management Plan based on the results of the biological and physico-chemical analysis carried out on the Rio Chichico.

CHAPTER 1

LITERATURE REVIEW

4.1 Aquatic biota

In its most common usage, the term biotic refers to all the species of plants, animals and other organisms that occupy a given area. For example, European biotic is used to refer to the list of species inhabiting that territory. The term biotic can be broken down into flora and fauna, according to the limits established in Botany and Zoology.

The concept can be extended to designate the repertoire of species in a soil compartment, the rhizosphere or the bottom of an aquatic ecosystem (prezi, 2005).

4.1.1 Macroinvertebrates

Carrera & Fierro (2005), mention that aquatic macroinvertebrates are bugs that can be seen with the naked eye and provide excellent signals about water quality. They are called macro because they are large (measuring between 2 mm and 30 centimetres), invertebrates because they do not have bones and aquatic because they live in places with fresh water (estuaries, rivers, lakes and lagoons) (p. 28).

These animals provide excellent signals about water quality, and by using them in monitoring, one can clearly understand the state of the water: some of them require good quality water to survive; others, however, resist, thrive and thrive in polluted water. For example, stoneflies live only in very clean water and disappear when the water is polluted. This is not the case for some larvae or maggots of other flies that resist pollution and thrive in dirty water.

As these organisms grow, they transform into flies that cause diseases such as malaria, malaria and Chagas disease.

Macroinvertebrates include insect larvae such as mosquitoes, damselflies, dragonflies or helicopterflies, bugs or chinch bugs, water dogs or alder flies.

Carrera Reyes & Fierro Peralbo (2005), state that macroinvertebrates provide excellent signals about water quality, and by using them in monitoring, the state of the water can be clearly understood: some of them require good quality water to survive; others, on the other hand, resist, grow and abound when there is pollution.

4.1.2 Benthic macroinvertebrates as indicators of water quality

Figueroa *et al* (2005) mention that it is common knowledge that organisms have been used for biological monitoring and water quality determination, however not all organisms have the advantages that benthic macroinvertebrates have. The advantages that these organisms provide in biological monitoring are outlined below:

• They allow for comparative studies because they are found in all the aquatic environments.

• Provides information on effects and disturbances that a body of water suffers, through an analysis of the pollution. This is mainly due to its sedentary nature.

• They have technical advantages, since both sampling and data analysis, it can be done in a simple and straightforward way.

• The taxonomy of the groups is now well studied and is you can find fact sheets and illustrations with which you can identify an organism.

The existence of numerous methods for data analysis, such as biotic indices and biodiversity, which are used in biomonitoring.

Biological methods were already used at the beginning of the century to determine the water quality of a river, but it was not until the 1950s that this type of study became more important (Figueroa *et al*, 2005).

4.1.3 Main Taxonomic Groups of Aquatic Macroinvertebrates

a) Order Ephemeroptera

According to Barber (2008) the Ephemeroptera (Ephemeroptera) are an order of pterygoth insects, commonly known as ephemeroptera, ephemeroptera, mayflies or "mayflies" in English. It is the oldest order of winged insects in existence today. There are 42 known families and more than 3,000 species inhabiting all biogeographical regions except Antarctica and some remote oceanic islands. The greatest diversity of genera belonging to this order is found in the Neotropical region (tropical areas of the American continent).

Ephemeroptera are an ancient lineage of insects, dating back to the Late Carborniferous or Early Permian periods, characterised by the fact that they spend most of their lives in the instar stage, in which up to 50 instars have been identified in some species (Barber, 2008).

As a main characteristic of the order Ephemeroptera, its nymphal larvae moult through a subimago stage, before developing into adults (imago). For this reason, many studies have been carried out in the South American region, resulting in the existence of a series of taxonomic keys for the identification of this order, facilitating studies, and improving the knowledge of species that are still to be known and studied on the South American continent (Dominguez, 2009).

b) Order Plecoptera

The earliest recorded fossil of Plecoptera dates from the Early Permian, i.e., 263-258 million years ago, and these organisms are phylogenetically related to the order Neoptera. Plecoptera are found in rivers where currents are strong,

however, some of their species have adapted to oligotrophic lentic environments, temporary systems, or deep lakes (Gutierrez, 2010).
These aquatic organisms are insects with incomplete metamorphosis (hemimetabolous), i.e. they only go through three stages of development: egg, nymph and adult. And before their transformation to the adult stage, the mature nymphs crawl on rocks, logs or any other substrate to get out of the water. In other cases, the nymphs climb several metres up trees (Gutierrez, 2010).
Gutierrez (2010) indicates that in feeding Plecoptera nymphs can be detritivorous, carnivorous or herbivorous; the type of feeding will depend on the stage of development, the species and even the time of day. Adults of most species do not feed, however, some may drink sugary substances or eat solid foods such as leaf litter, bquenes, fungal hyphae and pollen.
Plecoptera, like most aquatic insects, have a decisive influence on the aquatic system, as their existence is fundamental to the flow of energy and the recycling of nutrients (Gutierrez, 2010).
The most relevant particularity of this group of organisms is perhaps their response to changes in the environment, due to their high sensitivity, which makes them excellent indicators of water quality (Gutierrez, 2010).

c) Order Odonata

According to Barber (2008), the name Odonata comes from the Greek "odon" meaning tooth, referring to their strong imindibulae. Odonata, commonly known as dragonflies in their adult stage, are very brightly coloured and due to their ease of observation they have been given many names such as: damselflies, gallegos, pipilachas among others.
These insects have large compound eyes and a thorax that supports the four membranous wings that emerge from the thorax. Their abdomen is elongated and slender. It is important to know that the nymphs of this order are aquatic, in addition to possessing a precise mouthparts to hunt their prey, as they are predators (Ramirez, 2010).
Dommguez (2009) indicates that in the Order Odonata 5700 species are known worldwide, in South America the country with the most known species is Brazil, followed by Venezuela and Peru; however, there are large uninvestigated areas so that the odonate fauna is poorly known. New species are being discovered day by day, especially of the suborder Zygoptera, for this reason their systematics is not yet firmly established.
The larvae of this order grow through several larval stages and the number of moults depends on temperature and food availability.
They can feed on aquatic invertebrates, insects, other odonates and are even capable of feeding on tadpoles and small fish, making them predators in the

freshwater aquaculture system (Dommguez, 2009).
This order is divided into two sub-orders:

Suborder Anisoptera

The order Anisoptera differs from Zygoptera in that they are robust, broad and end in an anal pyramid. Their head is thinner than the width of their body (Ramirez, 2010).

- **Suborder Zygoptera**

The order Zygoptera is characterised by an elongated and slender body, with three caudal gills at its termination. The head of this order is larger than the width of its body (Ramirez, 2010).

d) Order Hem^ptera

Dommguez (2009) indicates that the Iremipte'ros or also called Heteroptera are divided into two types: Gerromorpha, which are those that live on the surface of the water, and Nepomorpha, which are those that live below the surface of the water. (pag. 167-169).

e) Order Coleoptera

According to Dommguez (2009) in the work Macroinvertebrados Bentonicos sudamericanos indicate that the Coleoptera are the most numerous group that is known, since it includes more than 350,000 species, distributed in 170 families and 4 suborders. Most Coleoptera are terrestrial, but there are 10,000 representative species that are aquatic in some stage of development; they are found in all inland waters, with the exception of heavily polluted lakes. (pp. 411-412).

f) Order Trichoptera

Springer (2010) mentions that Trichoptera (caddisflies) are holometabolous insects, resembling Lepidoptera, and in the adult stage are similar to small moths.

Their wings are covered with hairs instead of scales, hence the name (trichos: hairs; ptera: wings) (pp. 151-152).

Holzenthal (2007) indicates that this group is among the most important and diverse of all aquatic taxa. Larvae are vital within the trophic chain developed in the aquatic habitat, and their presence as abundance is used in the biological assessment of water quality. The morphology of this taxonomic group is determined according to three stages: adults, larvae and pupae. Trichoptera has 45 families distributed in 600 genera and approximately 13000 species.

g) Order Lepidoptera

Dommguez (2009) mentions that this order appeared millions of years ago, colonising all types of habitat, although most species are terrestrial, there are also species that are aquatic and their larval stages develop in water, and there

are even species such as (*Acentria*) whose adult stage is also aquatic.
In general, this group of organisms is not considered to be part of the aquatic community, which is why they are continually excluded in studies of the ecology of insects related to this type of environment. However, some species related to aquatic plants have now been claimed to be of interest (Dommguez, 2009).
According to Domrnguez (2009), because the order Lepidoptera is herbivorous, it is an important controller of aquatic weeds; however, they can be detrimental when they attack crops such as rice. Thus, within the order there are totally aquatic species, which develop all their stages in water, and semi-aquatic species that are related to submerged or emergent vegetation.
Existing information on this order in South America is fragmentary, with literature on aquatic larvae and detailing the most common genera in the region (Domrnguez, 2009).

h) Order D^ptera

Domrnguez (2009) indicates that the representatives of this order are holometabolous, recognisable by their inconspicuous colours and by having a pair of membranous wings. Within the order Diptera there are 153,000 species, with more than 158 families, of which 126, with 29,700 species in the Neotropical Region. The D^ptera, although known for 30 years, are difficult to identify due to the lack of modern synoptic treatments, the lack of collections, and the fact that some species have not even been described. Larvae are not known for more than 10% of the species. Therefore, their investigation is relevant since the different families of Diptera are important in veterinary medicine, as ecological and agronomic indicators (Domrnguez, 2009).

i) Phylum Platyhelminthes: Order Tricladida

According to Roldan & Ramirez (2008), within the Phylum Platyhelminthesse, we find the class Turbellaria, whose organisms are free-living. The class Turbellaria includes the Order Tricladida, which corresponds to the group of Planaria, they are slender and flat and can measure up to 30 ms. This group of organisms can live in water on branches, leaves, fallen trunks, in lentic and lotic waters; they live in oxygenated water but some species can live in waters that are highly contaminated by organic matter. Planarians are predators, but can also feed on dead animals; their main competitors are other predators such as Anelids, Nematodes and some Crustaceans (Roldan & Ramirez, 2008).

j) Phylum Mollusca: Class Gasteropoda

Gasteropods are organisms that have a spiral-shaped shell, ranging in size from 2 to 70 mm. They are found in waters with calcium carbonate, as this facilitates the construction of their shells. These organisms are associated with calm and shallow waters, where there is abundant aquatic vegetation and decomposing organic matter, being persistent in waters with hardness and high alkalinity (Roldan & Ramirez, 2008).

k) Phyllum Anelida

According to Roldan & Ramirez (2008) Anelids constitute a large group of organisms, but within the study of freshwater systems in the Neotropics, two classes are considered:

l) Oligochaeta and Hirudinea

The Oligochaeta class, ranging in size from 1 to 30 mm, are recognised as having the same structure as terrestrial earthworms. They feed on diatoms, filamentous algae and animal and plant detritus. These aquatic organisms are indicators of pollution, as they inhabit waters with high organic matter and low oxygen levels *(Roldan & Ramirez, 2008).*

Roldan & Ramirez (2008) indicate that the family Tubificidae is the most recognised, as it includes the genus Tubifex, which proliferates and develops in waters with high degrees of contamination, and can reach an abundance of 40,000 individuals per square metre.

Class Hirudinea includes the leeches, whose group of organisms can measure between 5 and 45 mm. Their representative characteristics are the suckers they possess, one at the front and one at the back, which they use to attach themselves or to move in water. They live in still water, ponds, reservoirs, attached to any substrate around them. They are predators and feed on snails, insects, water worms and some small macroinvertebrates (Roldan & Ramirez, 2008).

4.2 Water

According to BARBA (2005) water is one of the most widespread and abundant substances on planet earth. It is an integral part of most living things, both animal and plant, and is present in many minerals. P1

4.2.1Water Contamination

According to GALLEGO (2000) cited in (Clara, 2005) :

Pollution is the action and effect of introducing substances or forms of energy into, or inducing conditions in, water which, directly or indirectly, result in a detrimental alteration of its quality in relation to subsequent uses or its ecological function. p. 8

4.2.2 Water alterations

a) Physical alterations of water

Table 1 Physical alterations of water

Physical alterations	Characteristics and contamination indicating

Colour	Unpolluted water is usually slightly reddish, brownish, yellowish or greenish in colour, mainly due to the humic, ferric compounds or green pigments of the algae it contains. Polluted waters can have very different colours but, in general, no clear relationship between colour and type of pollution can be established.
Smell and taste	Chemical compounds present in the water such as phenols, various hydrocarbons, chlorine, decomposing organic matter or essences released by various algae or fungi can give very strong odours and flavours to the water, even if they are in very small concentrations. Salts or minerals give salty or metallic tastes, sometimes without any odour.
Temperature	The increase in temperature decreases the solubility of gases (oxygen) and generally increases the solubility of salts. It increases the speed of metabolic reactions, accelerating putrefaction. The optimal temperature of drinking water is between 10 and 14°C. Continue...
	Nuclear power plants, thermal power plants and other industries contribute to thermal water pollution, sometimes significantly.
Suspended materials	Particles such as clays, silt and others, even if they are not dissolved, are carried by the water in two ways: in stable suspension (colloidal solutions); or in suspension which only lasts as long as the movement of the water carries them along. Those suspended colloidally will only precipitate after they have undergone coagulation or flocculation.
Radioactivity	Natural waters have some radioactivity values, mainly due to isotopes of K. Some human activities can contaminate water with radioactive isotopes.
Foams	Detergents produce foams and add phosphate to the water (eutrophication). They greatly diminish the self-purifying power of rivers by hindering bacterial activity. They also interfere with the flocculation and sedimentation processes in sewage treatment

	plants.
Conductivity	Pure water has a very low electrical conductivity. Natural water has ions in solution and its conductivity is higher and proportional to the amount and characteristics of these electrolytes. This is why conductivity values are used as an approximate index of solute concentration. As the temperature changes the conductivity, measurements should be made at 20°C.

Source: World Health Organization (2012)

b) Chemical alterations of water

Table 2 Water Quality Alterations

Chemical alterations	Pollution indicating
pH	Natural waters can have an acidic pH because of dissolved CO2 from the atmosphere or from living organisms; because of sulphuric acid from some minerals, because of dissolved humic acids from the mulch in the soil. The main basic substance in natural water is calcium carbonate which can react with CO2 to form a carbonate/bicarbonate buffer system. Water polluted by mining or industrial discharges can have a very acidic pH. The pH has a great influence on the chemical processes that take place in the water, the action of flocculants, purification treatments, etc.
Dissolved oxygen (DO)	Clean surface waters are usually saturated with oxygen, which is essential for life. Low dissolved oxygen levels indicate contamination with organic matter, poor water quality and an inability to support certain forms of life.
Biodegradable organic matter: Biochemical Oxygen Demand (BOD5)	BOD5 is the amount of dissolved oxygen required by micro-organisms for aerobic oxidation of biodegradable organic matter in water. It is measured after five days. Its value gives an idea of the
	The quality of the water from the point of view of the organic matter present and allows to predict how much oxygen will be necessary for

	the purification of this water and to check the efficiency of the purification treatment in a plant.
Oxidisable materials: Oxidisable materials: Oxygen Qrnamic Demand (COD)	It is the amount of oxygen needed to oxidise the materials contained in the water with a chemical oxidant (usually potassium dichromate in an acid medium). It is determined in three hours and, in most cases, has a good relation to the BOD and is therefore very useful as it does not require the five dins of the BOD. However, COD does not differentiate between biodegradable and non-biodegradable matter and does not provide information on the degradation rate under natural conditions.
Total nitrogen	Several nitrogen compounds are essential nutrients. Their presence in water in excess causes eutrophication. Nitrogen occurs in many different chemical forms in natural and polluted waters. In the usual analyses, the NTK (total nitrogen Kendahl), which includes organic and ammoniacal nitrogen, is usually determined. The nitrate and nitrite content is given separately.
Total phosphorus	Phosphorus, like nitrogen, is an essential nutrient for life. Its excess in water causes eutrophication.
	Total phosphorus includes different compounds such as various orthophosphates, polyphosphates and organic phosphorus . determination is made by converting all of them into orthophosphates which are used in the production of the Continued... determined by chemical analysis.
Anions: Chlorides Nitrates Nitrites Phosphates	indicate salinity indicate agri'cola contamination indicate bacteriological activity

Sulphides Cyanides Fluorides	indicate detergents and fertilisers indicate anaerobic bacteriological action (sewage, etc.) indicatecontaminationoforigin industrial In some cases they are added to water for the prevention of caries, although this is a controversial practice.
Cations: sodium calcium and magnesium magnesium ammonium heavy metals	Indicates salinity They are related to water hardness. Contamination with fertilisers and faeces Of very harmful effects ; it is bioaccumulate in the trophic chain; (these are discussed in detail in the relevant chapter)
Organic compounds	Oils and fats from food waste from industrial processes (automobiles, lubricants, etc.) are difficult to
	metabolised by bacteria and float in the water forming films that harm living organisms. Phenols can be in water as a result of industrial pollution and when they react with chlorine added as a disinfectant they form chlorophenols which are a serious problem because they give water a very bad smell and taste. Contamination with pesticides, petroleum and other hydrocarbons is discussed in detail in the relevant chapters.

Source: World Health Organization (2012)

4.2.3 Physico-chemical characteristics of water

a) F^sicos

- **pH** - According to NOGALES (2005). The determination of pH is very

The quality of irrigation water is important, which must withstand ranges of 6.5 to 7.5 in order to ensure that crops are fit for human consumption and not sources of contamination.

- **Temperature** - According to NOGALES (2005). Temperature is a

physical factor that determines the solubility of gases and minerals; significantly influences the biological processes of respiration, growth of organisms and decomposition of organic matter.

- **Electrical Conductivity -** According to NOGALES (2005) "This

The parameter is determined by the concentration of dissolved ions in the water. Crops can tolerate different concentrations of ions and these determine the production and quality of products.

- **Turbidity - and according to** NOGALES (2005) Turbidity in the medium

The natural sediment can be organic, produced by algae and organic matter in suspension; and inorganic, consisting of particles of different sizes in suspension, such as clays, especially introduced in the course of the channel, produced by erosion of the watercourse.

- **Total Dissolved Solids -** NOGALES (2005). Says that the

Solids analyses are important in the control of physical and biological processes. STD has a special significance because many waters contain unusual amounts of dissolved inorganic salts, this parameter is related to conductivity and like STD.

b) Qrnmicos

- **Alkalinity -** According to NOGALES (2005) "It is the consumption of acid by the basic species present in natural waters. Alkalinity is mainly due to the presence of salts of weak acids and strong bases".
- **Hardness -** According to NOGALES (2005). Hard water is defined as

those that generally require considerable amounts of soap to produce foam; the hardness is caused by divalent metal cations that are able to react with the soap to form precipitates; it does not affect human health, but it does affect vegetables by hardening them.

- **Chlorides -** According to GUERRERO (2005), chloride is one of the most important

The main inorganic anions in natural and waste water, which significantly impair plant growth, chloride levels are very important for the selection of water supplies for drinking and irrigation.

- According to GUERRERO (2005) "El

water rich in organic matter can cause pollution. Organic matter is mainly composed of proteins, carbohydrates, animal fats, which can be carried along the irrigation canal.

- **Dissolved oxygen -** Dissolved oxygen is derived from the

photosynthetic activities of plant flora and atmosphere-water exchange, the levels of dissolved oxygen in natural waters or waste waters depend on the degree of physico-chemical and biochemical activity. Dissolved oxygen analysis is a fundamental test in the process of monitoring contaminated water. The dissolved oxygen is used to calculate the oxygenation index. GUERRERO (2005).

4.3 Environmental Management Plan

The Environmental Management Plan (EMP) is a management instrument whose purpose is to serve as a guide for programmes, procedures, practices and actions aimed at preventing, minimising, mitigating and controlling the environmental impacts and risks generated by different activities (Biosfera Gestion Ambiental, 2009).

The Unified Text of Secondary Environmental Legislation (TULSMA), in its Book VI, establishes the actions required to prevent, mitigate, control, compensate and correct the possible negative environmental effects or impacts caused in the development of a project, work or activity; it also includes follow-up, evaluation and monitoring plans and contingency plans. The content of the plan may be regulated differently in each country.

4.4 Legal framework

The current Constitution of the Republic of Ecuador was published by Official Register No. 449 of 20 October 2008, contains the principles, rights and freedoms of those who make up the Ecuadorian society and constitutes the apex of the legal structure of the Ecuadorian State.

In Title II, Chapter II, on rights, Chapter II, on the rights of the good life, Section I, water and food, Articles 12, 71 and 72 maintain the following as primary duties of the state

rights to water and nature or Pacha Mama,

Art.12.- The human right to water is fundamental and inalienable. Water constitutes a strategic national patrimony for public use, inalienable, imprescriptible, unseizable and essential for life.

Art.71.- Nature or Pacha Mama, where life is reproduced and realised, has the right to full respect for its existence and the maintenance and regeneration of its vital cycles, structure, functions and evolutionary processes.

Art. 72.- Nature has the right to restoration. This restoration shall be independent of the obligation of the State and natural or legal persons to compensate individuals and collectives that depend on the affected natural systems.

In cases of serious or permanent environmental impact, including those caused by the exploitation of non-renewable natural resources, the State shall establish the most effective mechanisms to achieve restoration, and adopt appropriate measures to eliminate or mitigate harmful environmental consequences.

The Constitution of the Republic of Ecuador of 2008 through the second chapter on the Rights of Good Living (Section one of Water and Food) states in Article No. 12 "The human right to water is fundamental and inalienable. Water constitutes national strategic patrimony of public use, inalienable, imprescriptible, unseizable and essential for life".

Article No. 74 of the Seventh Chapter on Rights of Nature states that "Individuals, communities, peoples and nationalities shall have the right to benefit from the environment and the natural wealth that allow them to live well. Environmental services shall not be subject to appropriation; their production, provision, use and exploitation shall be regulated by the State".

In Chapter VI of the Water Codification Law through Article No. 22 it mentions:

"All water pollution affecting human health or the development of flora or fauna shall be prohibited. The National Council of Water Resources, in collaboration with the Ministry of Public Health and other State entities, shall apply the policy that allows compliance with this provision. A popular action is granted to denounce facts related to water pollution. The complaint shall be submitted to the Ombudsman's Office".

Chapter VI of the Environmental Pollution Prevention and Control Act in Articles 16 and 18 states the following:

Art. 16.- It is forbidden to discharge wastewater containing pollutants that are harmful to human health, fauna and property into sewage systems, or into streams, ditches, rivers, natural or artificial lakes, or into marine waters, or to infiltrate into land, without complying with the corresponding technical standards and regulations.

Article 18 - The Ministry of Health shall determine the degree of treatment to be given to the liquid waste to be discharged into the receiving body, whatever its origin.

Article 12 of the Environmental Sanitation Title I of the Environmental Pollution Prevention and Control Act states that "No person shall dispose of solid, liquid or gaseous wastes into the air, soil or water without prior treatment to render them harmless to health".

On 31 March 2003, the Unified Text of Environmental Legislation was issued by Executive Decree 3516, published in Official Gazette No. 2, with the following content:

Preliminary Title: Basic Environmental Policies of Ecuador.

Book I: The Environmental Authority.

Book II: Environmental Management.

Book III: Forestry Regime.

Book IV: On Biodiversity.

Book V: Coastal Resources.

Book VI: Environmental Quality.

Book VII: Special Regime: Galapagos.

Book VIII: The Institute for Amazonian Regional Eco-development.
Book IX: On the System of Fees or Charges for the Services rendered by the Ministry of the Environment and for the Use and Exploitation of National Assets under its Charge and Protection (www.ambiente.gov.ec).

Book VI Annex 1 states that this technical standard determines or establishes:

Permissible limits, provisions and prohibitions for discharges into water bodies or sewage systems; water quality criteria for different water uses and methods and procedures for determining the presence of pollutants in water.

4.5 Conceptual Framework

Environment: It is the global system made up of natural and artificial elements of a physical, chemical, biological, sociocultural nature and their interrelationships, in permanent modification by human or natural action that governs or conditions the existence or development of life.

Wastewater: Water resulting from a process or productive activity whose quality has been degraded due to the incorporation of pollutants.

Wastewater: Aqueous waste resulting from the disposal or use of water in any activity that may cause contamination.

Bio-indicator of water quality: A bio-indicator is an indicator consisting of a plant, fungus or animal species; or formed by a group of species (eco-sociological group) or plant grouping whose presence (or state) gives information on certain ecological characteristics (physico-qrnmic, micro-climatic, biological and functional) of the environment, or on the impact of certain practices on the environment. They are mainly used for environmental assessment (monitoring the state of the environment, or the effectiveness of compensatory or restorative measures).

Biosafety: Evaluation, regulation and management of biological risk, through its reduction or elimination, to preserve the qualities and survival of human life and other organisms in the environment.

Biomonitoring: Biomonitoring is a set of techniques based on the reaction and sensitivity of different living organisms to various substances (nutrients/pollutants) present in an ecosystem.

It is the evaluation of the deleterious effects of a toxic substance on various organisms. These organisms are used as biological indicators of the toxicity of a compound which is measured by different biological indicators (alterations in development, in vital functions, etc.).

A set of techniques associated with environmental quality assessment using living organisms (Col Kent 2005).

Climate: Set of atmospheric conditions (temperature, humidity, cloud cover, rainfall, sunshine, direction and wind speed) that dominate and alternate continuously in a given locality.

Contamination: Undesirable change of physical, chemical and biological properties that can cause negative effects on the different components of the environment.

Pollutant: Chemical, biological or radiological substance, in any of its physical states and forms, which when incorporated or found above its normal concentrations in the atmosphere, water, soil, fauna or any natural element alters and changes its composition and natural condition.

Freshwater aquaculture: Plants and animals that carry out all their vital functions in freshwater.

Environmental damage: Negative or detrimental action exerted by a factor or factors outside the environment.

Biological degradation: Negative action exerted on a given ecosystem that partially or totally changes its characteristics.

Biochemical Oxygen Demand: Indirect measure of the amount of oxygen consumed by micro-organisms during the biological degradation of organic matter.

Discharge: The disposal or addition of waste or residuals to a receiving medium.

Waste: Materials resulting from a production or research process that cannot be modified for production, transformation or consumption purposes and that are to be disposed of.

Effluent: Liquid waste, treated or untreated, originating from an industrial process or social activity and generally disposed of in soils or various surface or groundwater bodies.

Environmental impact assessment: Compilation and evaluation of reports on the physical, ecological, economic and social characteristics of a specific area or region, as well as the plans and projects to be implemented there, in order to minimise negative impacts on the environment.

Liquid effluents: Liquid effluents are essentially the water supply of a population, after having been impurified by various uses. From the point of view of their origin, they are the result of the combination of the liquids or wastes carried by water from dwellings, institutions and commercial and industrial establishments, plus any groundwater, surface water or rainwater that may be added.

All these waters affect in some way the normal life of their corresponding receiving bodies. When this effect is sufficient to render them unsuitable for further use, they are said to be polluted.

Fertiliser: Fertiliser is a substance that is added to the soil to supply those elements that are required for plant nutrition. A fertiliser or carrier material is a substance that contains one or more of the essential elements for plants. Complete fertilisers contain the three major plant nutrients: nitrogen, phosphorus and potassium.

Environmental impact: Impact on the environment caused by anthropogenic action or by an element external to the environment, which generates notable consequences in the environment.

Tolerance limit: Maximum concentration of a harmful substance that can be tolerated in the environment for a prolonged period of time without harm. Amount of a chemical substance in food which is considered safe for humans and animals.

Environmental management plan: An environmental management plan is a detailed plan that establishes the actions required to prevent, mitigate, control, compensate and correct the possible negative environmental effects or impacts caused by the development of a project, work or activity; it also includes follow-up, evaluation, monitoring and contingency plans.

Rehabilitation: A set of actions aimed at re-establishing the original altered conditions of an ecosystem or one of its elements.

Residues: Waste substances from a physiological production process which, when directly incorporated into the environment, alter its state, usually in form.

Sustainability: The ability of a system to develop with its own resources in such a way that its functioning does not depend on external sources, although this does not mean that these are not considered.

Taxonomy: Taxonomy is the science that studies the principles, methods and purposes of classification. The term is used especially in biology to refer to a mainly ordered and hierarchical classification of living things.

TULSMA: Unified Text of Secondary Environmental Legislation.

CHAPTER 2

MATERIALS AND METHODS

5.1 Materials

The following equipment, tools and instruments were used in this investigation:

5.1.1 Equipment

- GPS Brand: Trimblejuno SB ID No. jup66410
- Camera Brand: Konica NLT-7D N° 9896R6R7R76.
- Microscope

5.1.2 Tools

- Rubber boots
- Tweezers
- Nylon netting type D 250 inn
- Gloves
- Small plastic bottles 10ml.
- Alcoholantiseptic75%.
- Field notebook
- Magnifying glass
- Apron

5.1.3 Instruments

- Political location of the city of Puyo Scale: 1:260:000

5.2 Methods

5.2.1 Location of the study area

The study was carried out in the Rto Chichico, located in the parish of Tarqui, rural area of the canton of Puyo, approximately 10 minutes from the city of Puyo; the study area has been delimited 100 metres from the river at four points for the collection of samples in order to analyse the situation of the water resource for its use (GAD Tarqui).

5.2.2 Political Location

The urban parish of Puyo has an extension of 104 km^2 . As we have seen when we speak of Puyo as the cantonal capital city of the canton of Pastaza and capital of the province, its urban perimeter is becoming more and more extensive, which has forced the municipality to extend its urban perimeter, its current boundaries are:

- North: With the parishes of Puyo and Veracruz.
- South: With the parish of Madre Tierra.
- East: With the parishes of Pomona and Madre Tierra.
- West: With the parishes Madre Tierra and Shell.

Figure 1 Political Location

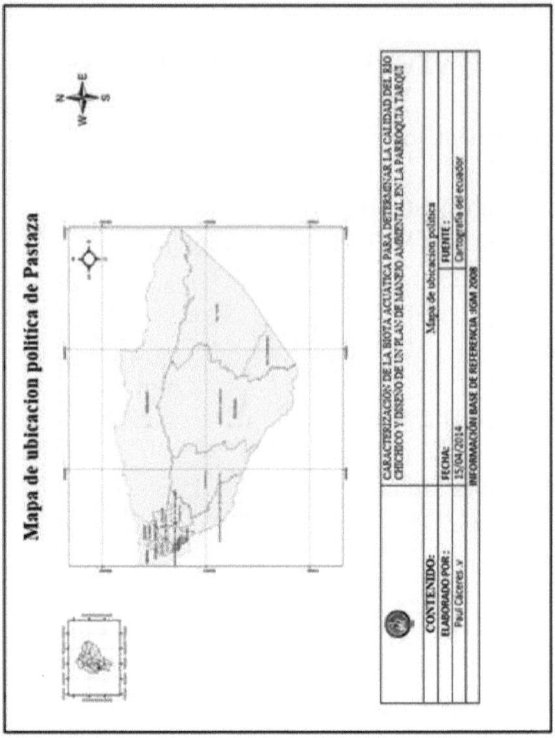

Prepared by: The Author

5.2.3 Geographical Location

Puyo is strategically located in the centre of the Ecuadorian Amazon, 4 hours from Quito and 7 hours from Guayaquil.

Figure 2 Geographical Location

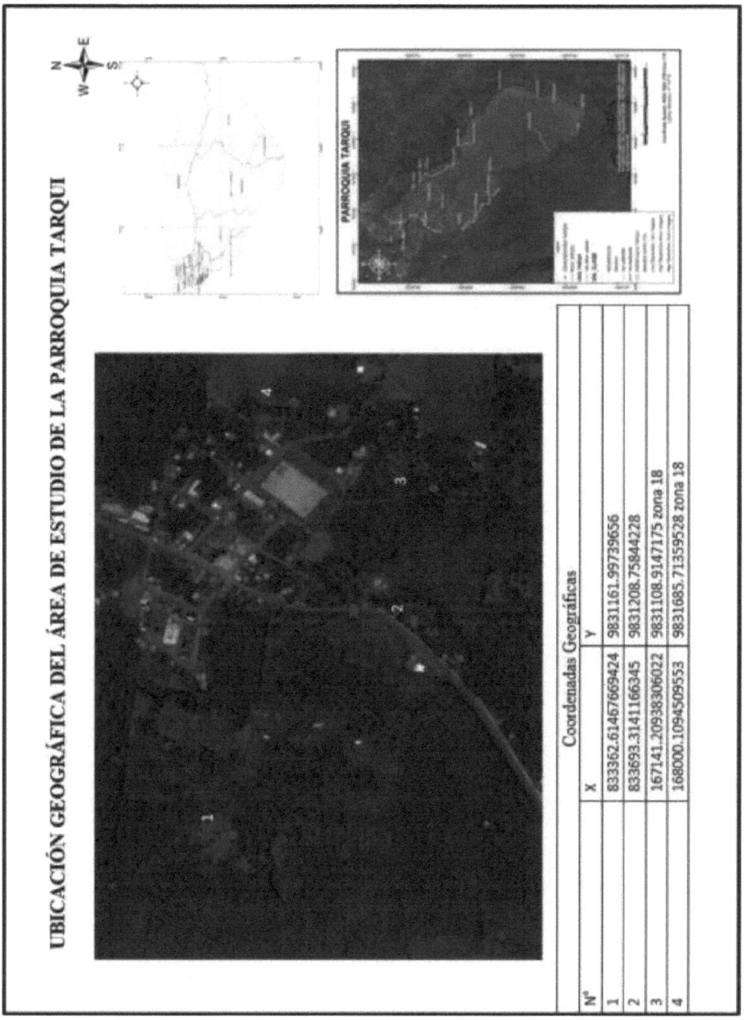

Coordenadas Geográficas		
N°	X	Y
1	833362.61467669424	9831161.99739656
2	833693.3141166345	9831208.75844228
3	167141.20938306022	9831108.9147175 zona 18
4	168000.1094509553	9831685.71359528 zona 18

Prepared by: The Author

5.3 Biophysical and climatic aspects

5.3.1 Biophysical aspects:

a) Hydrography

The hydrographic system of the parish of Tarqui is comprised of the basins of the Puyo and Putuimi rivers. Other rivers of importance are the Salome, Chichico, Pindo Grande, Chingushimi, Rosario Yacu and the Palimbe, Chicocoyak and Paliaba estuaries (GAD Tarqui).

b) Geomorpholog^a

The area analysed includes part of the eastern sub-Andean foothills, the nearby piedmont with volcanic ash cover and the fluvial landscapes of the valleys with undifferentiated terraces.

The physiographic characteristics of the area are determined by the geological, tectonic and climatic events that occurred between the Upper Tertiary and Quaternary, as well as by the erosive agents that continue to act over time (GAD Tarqui).

c) Flora

Table 3 Flora of the city of Puyo

NAME COMMON	NAME CIEVHFICO	FAMILY
Mani de Monte	*Caryodendron rhinocense*	Euphorbiaceae
Drago	*Croton sp*	Euphorbiaceae
Pambil	*Iriartea deltoidea*	Arecaceae
Wild grapes	*Pouruma bicolor*	Ericaceae
Caimito	*Pouteria caimito*	Sapotaceae
Chontaduro	*Bactris gasipaes*	Arecaceae
Aji	*Capsicum chinense*	Solanaceae
Breadfruit	*Artocarpus altilis*	Moraceae
Mai'z	*Zea mays*	Poaceae
Cassava	*Manihot esculenta*	Euphorbiaceae
Sugar cane	*Saccharum officinarum*	Poaceae
Papaya	*Carica papaaya*	Caricaceae
Banana	*Musa paradisiaca*	Musaceae
Orito	*Musa acuminatta*	Musaceae
Chinese potato	*Colocasia esculenta*	Araceae
Sweet potato	*Ipomoea triloba*	Convolvulceae
Morete	*Mauritia flexuosa*	Araceae
Achote	*Miconia sp.*	Tucano

Prepared by: The Author

d) Fauna

Table 4 Fauna of the area

NAME	NAME	FAMILY
Squirrel	*Sciurus ingiventris*	Scyuridae
Danta	*Tapirus Pinchaque*	Cathartidae
Guatusa	*Dasyprocta fuliginosa*	Dasyproctidae
Rabbit	*Silvilagus brasiliesis*	Leporidae
Little bat	*Platyrrhinus lineatus*	Vespertilionidae

Long-eared bat	*Tonatia silvicola*	Vespertilionidae
Large Raposa	*Didelphis marsupialis*	Mustelidae

Prepared by: The Author

5.3.2 Climate aspects:

a) Climatic Floors.

The climate of the region is warm and humid due to the permanent rainfall, which reaches 4.000 mm/year. It varies according to the climatic zones and the altitude of the populations. The average luminosity is 1003 hours/day, with very low radiation due to the permanent cloud cover (Plan vial de la Provincia de Pastaza 2013 - 2025).

The variety of climate types includes:

- **Tropical: Very Humid Temperate Warm:** Corresponds to the

The average annual temperatures are between 14°C and 18°C, and the average annual rainfall is between 2,500 and 3,000 mm. This unit is located in the extreme east of the province and occupies 0.37% of the total area (Plan vial de la Provincia de Pastaza 2013 - 2025).

- **Sub Tropical: Very Humid:** It is located between 700 and 1.200

The average annual temperature is between 16 and 20 °C, with an average annual rainfall of 2,000 to 4,000 mm. This unit constitutes 1.76 % of the total area (Plan vial de la Provincia de Pastaza 2013 - 2025).

- **Sub Tropical: Rainy:** This is a micro-climate located in the

This is a small portion of the entire province, approximately 3% of the basin, and is characterised by rainfall of more than 3,000 mm, with values of 4,000 to 5,000 mm, with an average temperature of 22 to 24 °C, close to the town of Puyo. (Plan vial de la Provincia de Pastaza 2013 - 2025).

- **Tropical: Humid:** It is typical of areas between 200 and 800 m above sea level,

where an average annual temperature of 22 and 26 °C and an average annual rainfall of 2,000 to 4,000 mm are recorded. This unit is located in the extreme east of the province, affecting 93.72% of the province (Pastaza Province Road Plan 2013 - 2025).

b) Climate

In order to describe the main climatic characteristics of the parish of Tarqui, the analysis of these conditions was taken from the INAHMI data of the year 2010, from the Puyo Meteorological Station; its conclusions are practically valid for all the geographical sectors, since the climatic conditions have a regional character and practically do not change over short distances (INAHMI).

c) Precipitation

Regarding the amount of rain that falls on the territory annually, as the climate of the region is mega-thermal rainy, the total rainfall is around 4442 mm, in accordance with its equatorial character, rainfall is abundant every month, with light maximums occurring in March, April, May and June, and minimums occurring more frequently in the months of July, August and September. The rainiest months have average values of 445.5 and the driest months have values of up to 273.8 mm.

The city is subject to heavy rainfall during most of the year, especially during the period between April and July. The average annual rainfall determined at the Tena station is 3,898.3 mm (INAHMI).

d) Temperature

The climate is tropical humid, equatorial type, i.e. always warm and humid. The average annual temperature of the region is 20,8°C, with a very low variability throughout the year. The absolute maximums are around 32,6°C and the absolute minimums are 12,1°C for a few days and hours of cold. (INAHMI).

e) Altitude range

The altitude range for Tarqui oscillates from 838 to 1.091 meters above sea level and presents a varied physiography ranging from Sub Andean reliefs, Amazonian foothills, and fluvial landscapes (INAHMI).

5.4 Type Of Research:

The research corresponds to a non-experimental design; it is based on field research, descriptive and documentary research; it consists of direct observation, by means of maps, registers, data collection, bibliographic revision, laboratory water analysis, biological indices to determine the quality of the water of the Chichico River.

Taking into consideration processes to be followed and applied, being of vital importance for the achievement of the proposed objectives.

a) Descriptive research - Descriptive research allows for the detailing and explanation of a natural or social problem or object, in order to determine characteristics of the observed problem.

The present investigation allowed to gather all the necessary information from a direct source, in this case the Chichico River, and from this to determine its alterations in the biotic factors.

b) Field research - Every research process needs to be analysed in situ for subsequent analysis in laboratories, so in the present study, this research was fundamental, since the characterisation of the aquatic biota was carried out in the Chichico River, and from this it allowed the establishment of permissible limits for water quality.

c) Documentary research: The documentary research allowed the establishment of all the theoretical foundations and the methods and techniques to be applied in the in situ monitoring, as well as the establishment of the environmental management proposal according to the studies carried out.

5.5 To determine the procedure for the sampling and identification of macroinvertebrates in four points of the Rio Chichico.

For the identification of macroinvertebrates at the study site, four sampling points were established:

5.5.1 Location of sampling sites

The sampling sites were georeferenced every 100m, with the help of a portable GPS Trimblejuno SB ID No. jup66410, a reconnaissance tour was carried out upstream along the Chichico River, where the following recommendations given by (Carrera and Fierro, 2001) were applied. (See Annex 1 Georeferenced points for macroinvertebrate sampling).

Taking into consideration and asking ourselves the following questions:

- (Does the shore have abundant vegetation?
- ^There are areas with a wide variety of animal species and plants?
- (Are there crops near the river?
- Are there livestock in the nearby watershed?
- (areas of the river channelled, dammed or diverted to irrigation?
- (The water is clear and clear?
- Do you have any strange smells?
- There is rubbish, plants or logs cutting off the flow of the water and creating pools?
- (Is solid or industrial waste dumped in the river?
- (Are there bank collapses?
- (The river has many streams, pools and rapids, one of which is a continuation of another?

5.5.2 Sample collection

Sampling was carried out using a Surber net of 30 x 30 cm surface area and 0.5 mm mesh opening, according to (Carrera and Fierro, 2001). The arc was placed on the bottom against the current and the material on the bottom inside the base or metal frame was removed with the hands for one minute; to do this, the arc was placed to one side of the net, so as not to block the water current with the body and prevent sediment from entering the net.

Once the sediment was collected, the collected samples were placed in a white tray for better visibility, where the macroinvertebrates were separated from the sediment and other particles.

5.5.3 Monitoring and control

In the field phase, four samples were taken, one for each sampling point along the Rto Chichico, the distance is defined every 100m; to identify the families of existing macroinvertebrates.

5.5.4 Sample Fixation

The macroinvertebrates collected were placed in 10 ml plastic bottles, and 70% alcohol was used to preserve the sample.

5.5.5 Labelling and transport of the sample

The containers were labelled with the following information: code, date, name of the person doing the sampling, coordinates and then taken to the laboratory of the Escuela Superior Politecnica de Chimborazo (ESPOCH) for identification.

5.5.6 Laboratory analysis

In the laboratory, the samples were separated and the macroinvertebrates were removed from the substrate with the help of a watchmaker's forceps and then placed in a Petri dish and taken to the microscope for identification, where physical characteristics were observed.

The macroinvertebrates collected were stored in test tubes with 70% alcohol. Specimens were identified using dichotomous keys used for neotropical aquatic entomofauna (Dommguez and Fernandez, 2009; Fernandez. and Dommguez, 2005; Manzo, 2005; Merritt and Cummins, 1988; Roldan, 1988; Salles, 2006).

Photo 1

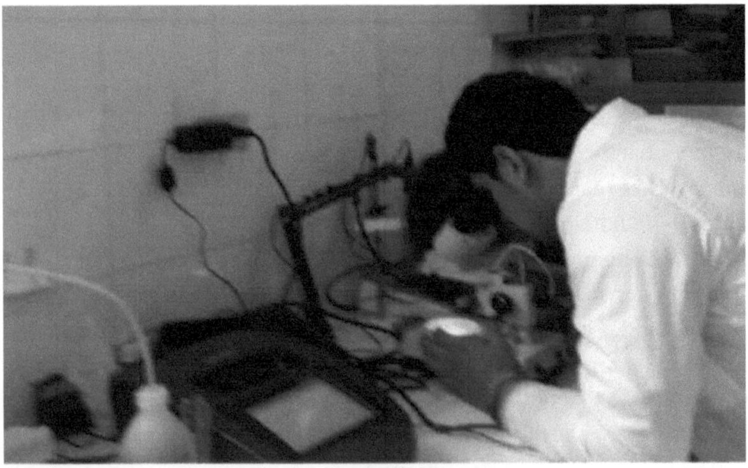

Identification of macroinvertebrates in the laboratory

5.5.7 Analysis of results

Once the macroinvertebrate families had been identified, the EPT index was calculated, which refers to the presence or absence of the orders Ephemeroptera,

Plecoptera and Trichoptera in a biological community. The species of these insect groups are sensitive to human disturbance (Carrera and Fierro, 2005).

5.5.8 EPT Index (Ephemeroptera, Plecoptera and Trichoptera)

This analysis was done using three groups of macroinvertebrates that are indicators of water quality because they are more sensitive to pollutants.

These groups are: Ephemeroptera or mayflies, Plecoptera or stoneflies and Trichoptera. Field sheets should be applied for EFA analysis.

One of these sheets was completed for each sampling area. Once the groups present in each area had been identified, the number of macroinvertebrates against the corresponding group was recorded in the **Abundance of Individuals** column of the Field Sheet. If any group did not correspond to any of the groups listed, the number of individuals was recorded against the **Other groups** row. All the numbers in the abundance of individuals column were added together and the result was recorded in the **Total** table (see annex table 1: Field sheet).

5.6 To evaluate the physical and chemical quality of the water of the Rio Chichico.

5.6.1 Selection of sampling points:

The sampling sites were georeferenced every 100m, with the help of a portable GPS Trimblejuno SB ID No. jup66410, we proceeded to a reconnaissance tour upstream by the Rto Chichico, where we applied as a reference the Ecuadorian Technical Standard NTE INEN 2 169:98.

5.6.2 Procedure for the collection, preservation and storage of samples for laboratory use.

a) Type of samples

- **Composite sample:**

Composite samples were obtained manually, four single samples were taken continuously, taken at different points, depending on the width and depth of the river, with the help of a container calibrated by litres, in order to meet all the conditions and specific requirements for the analysis in the laboratory.

b) Filling the container

For the analysis of the physico-chemical parameters, the gallons were filled completely and immediately covered in such a way that no air was present in the sample.

c) Types of sample containers

Polyethylene bottles with a capacity of one gallon were used to take samples for the analysis of the physical and chemical parameters of the waters of the Rio Chichico.

d) Identification and registration

For the identification and registration of the samples, a letterhead was prepared with the following information:

- location (and name) of the sampling site, with co-ordinates

and any relevant location information;

- details of the sampling point;
- date of collection;
- method of collection;
- time of collection;

name of the collector;

data collected in the field.

And then it was immediately attached to the gallon after the letterhead was filled in (see photo annex 6,7).

e) Transport of the sample

The containers (gallons) with the samples were placed in a cooler at a temperature of 4 - 5 °C and immediately sent to the laboratory of the Escuela Politecnica del Chimborazo (ESPOCH) for analysis (see annex photo 8).

f) Parameters for laboratory analysis

The physical-chemical parameters selected according to the technical criteria of the researcher, based on the Regulations of the Unified Text of Secondary Environmental Legislation, Book VI, Table N°01, on the use of water for human consumption and domestic use, this measure was considered, since the community uses the water resource for consumption in their homes, It is therefore of great importance to know the maximum permissible limits contained in the water in order to establish whether or not it complies with current regulations, and to determine the quality of the water that flows through the Chichico River, which will serve to establish corrective measures in the proposed Environmental Management Plan.

Table 5 Physico-chemical parameters to determine the water quality of the Rio Chichico.

ANALYSIS	PARAMETERS	UNIT
F^SICAL	DissolvedSolids Totals	mg/l
	Conductivity	^s/cm
CHEMICALS	Hydrogen Potential	
	BOD	mg/l
	COD	mg/l
	Dissolved oxygen	mg/l
	Phosphate	mg/l
	Nitrate	mg/l
	Nitrite	mg/l

	Sulphate	mg/l

Prepared by: The Author

5.7 Propose a Management Plan on the basis of the results of the biological and physico-chemical analysis carried out on the Rio Chichico.

In order to propose solutions to the possible environmental problems existing in the Rto Chichico, it is necessary to incorporate the Environmental Management Plan, with the objective of mitigating, preventing or controlling the possible impacts generated by the different human activities along the banks of the river area under analysis. For which we present the following structure of the proposed Environmental Management Plan.

1. Introduction
2. Target
3. Outreach
4. Proposed Environmental Management Plan
5. Environmental prevention and mitigation programme
6. Waste Management Programme (WMP)
7. Environmental training programme
8. Programme for the rehabilitation of affected areas
9. Environmental Monitoring Programme
10. Programme budget for the Environmental Management Plan.

5.7.1 Introduction

The content of this item is based on the problems and objectives developed in the present investigation, information that serves as a basis for the elaboration of the Environmental Management Plan for the recovery of the Rto Chichico.

5.7.2 Objective

The objectives will be framed according to the needs to prevent, control and mitigate possible negative impacts on the banks of the Rto Chichico.

5.7.3 Scope

It will describe the area and the actors that will be involved in the Environmental Management Plan.

5.7.4 Environmental Management Plan Proposal

This chapter details how the Environmental Management Plan is designed with its respective programmes for its proposed implementation, in order to recover the water quality of the Rto Chichico.

a) Environmental Prevention and Mitigation Programme: Based on the criterion that it is always better to prevent and minimise the occurrence of negative environmental and social impacts than to mitigate or correct them. The environmental mitigation measures are intended to avoid or reduce the adverse effects of the project or activity, whatever its phase of execution.

b) Waste Management Programme: Aimed at establishing criteria to identify, categorise, recycle, reuse, control and dispose of degradable and non-degradable, hazardous and non-hazardous, industrial and domestic waste to be generated, in compliance with environmental regulations and standards.

c) Environmental Training Programme: By identifying the minimum content necessary for people to carry out the tasks of good water resource management.

d) Environmental Rehabilitation Programme: This involves the recovery over time of the morphology and vegetation cover of the impacted areas.

e) Monitoring Programme: Focused on obtaining analytical information for: Follow-up related to the restoration of intervened and/or affected areas.

f) Budget programme for the Environmental Management Plan: Where the cost of each of the programmes to be implemented is broken down:

CHAPTER 3

RESULTS

6.1 Identify the presence of macroinvertebrates at four points in the Rio Chichico.

At point 1 of the Chichico River, 14 families were identified, distributed in 5 orders belonging to the insect class, species that were observed during the field survey in the months of May, June and July 2014, information that is indicated in Table 1 (see photo annex 9,10).

Table 1 Results of sampling point 01

KINGDOM	PHYLUM	CLASS	ORDER	FAMILY	ABUNDANCIA		RATE OF EPT PRESENTS
					N	%	
Animalia	Arthropoda	Insecta	Ephemeroptera	Baetidae	20	21,74	20
Animalia	Arthropoda	Insecta	Diptera	Ceratopogonidae	4	4,35	0
Animalia	Arthropoda	Insecta	Diptera	Chironomide and	5	5,43	0
Animalia	Arthropoda	Insecta	Neuroptera	Corydalidae	6	6,52	0
Animalia	Arthropoda	Insecta	Coleoptera	Elmidae	8	8,70	0
				Euthyplociidae	2	2,17	2
				Glossosomatidee	1	1,09	1
Animalia	Arthropoda	Insecta	Trichoptera	Hydropsychidae	15	16,30	15
Animalia	Arthropoda	Insecta	Trichoptera	Leptoceridae	10	10,87	10
Animalia	Arthropoda	Insecta	Trichoptera	Leptohyphidae	5	5,43	5
				Oligochaeta	2	2,17	0
				Oligoneuridee	3	3,26	3
				Perlidae	5	5,43	5
				Philopotamidae	1	1,09	1
				Other groups	5	5,43	0
				TOTAL	92	100	62
				EPT TOTAL - TOTAL ABUNDANCE	TOTAL ABUNDANCE		62- 92 = 0,63 0,63 x 100 = 63%

Prepared by: The Author

In table 6, it is noted that according to the EFA index methodology, the water quality at sampling point N°01 is determined to be good, as its rating is 63%.

Graph 3

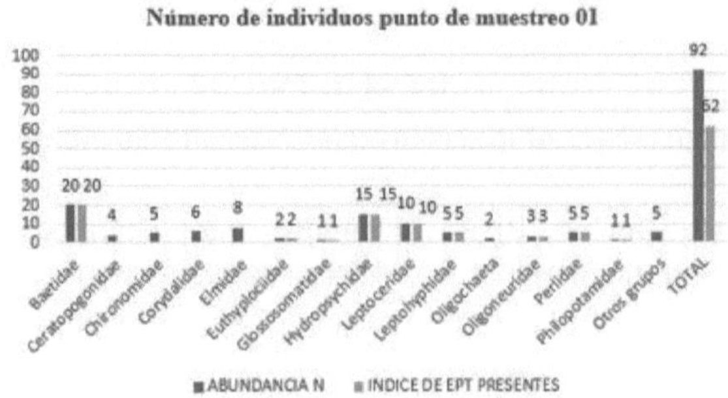

Prepared by: The Author

Interpretation: according to graph 3, 20 individuals corresponding to the Baetidae family have been identified. They are insects adapted to living in very different aquatic habitats, they are good swimmers, certain species are sensitive to contamination, but there are also species that are tolerant to moderate levels of organic pollution, so it can be determined, according to the number of specimens found in the Chichico River, that they have a low level of organic matter.

We also found 4 individuals corresponding to the family Ceratopogonidae, these insects belong to the group of d^ptera, they adapt to different environments, as we found a relatively low number, it cannot be taken as an indicator for the evaluation of the quality of the Rto Chichico.

In the case of the **Chironomidae** family**, 5 individuals were found, which are biological indicators that determine the presence of** oxygen, but due to their small number they cannot be taken as an indicator for the evaluation of the quality of the Chichico River.

Family Corydalidae, 6 specimens were found, this kind of insects were easy to identify because they presented large wings with protruding venation, the anterior ones were slightly larger than the posterior ones, three ocelli were observed on the head and filiform antennae, as their larvae are good indicators of water quality, good, but not high, quality was determined.

Family: Elimidae, 8 specimens were collected that live in streams and rivers, both larvae and adults were found at sampling point number one, these specimens were found in areas where plant remains accumulate, which they feed on, they need oxygenated water, so it is a specimen that can be associated with water quality, the specimens found were in low numbers so we can not speak of optimal water quality in the Rio Chichico.

Family: Euthyplociidae 2 specimens were found in the Rio Chichico, only two adults, grey, dark brownish, with a medium-light stripe on the abdomen, large wings, and a wingspan of 70 mm. As they are good indicators of clean water, the low number of specimens indicates that the water quality of the river is inadequate.

Family: Glossosomatidae 1 specimen was found in larvae in an area of stony substrate of the Rio Chichico, its physical characteristics observed were in the form of a coffee bean, flat bottom and two openings, these specimens are considered as indicators of water quality, so that only by finding a single specimen, it is considered that the water quality of the Rio Chichico is good.

Rto Chichico, not suitable, in agreement with the other specimens found.

These specimens were easily recognised by their ventral abdominal gills and fully sclerotised thorax. These families were found along the length of the river, and in the case of the Hydropsychidae, they are not considered to be water quality indicator groups. Their numbers are not indicative of high or low water quality in the Rio Chichico.

Family: Leptohyphidae 10 specimens, The Leptohyphidae specimens were described by their slightly flattened body, with wide mesothorax, their legs with wide femurs and in the first pair a transverse row of setae or spines, the individuals were found in the river in an area of leaf litter, being a family that adapts easily to different environments and water qualities, it is not a specimen that helps us to determine the optimum water quality of the Rio Chichico.

Family: Leptoceridae 5 specimens, Found specimens with long cases of varied composition, the third pair of legs were observed much longer, in the case of this family being very sensitive to contamination and indicators of good quality, due to the low number of specimens found, it is determined that there is not an optimal environment.

Family: Oligochaeta 2 specimens, this group was heterogeneous, this specimen adapts to different environments, they are an eminently detritivorous group, they are very abundant in waters rich with organic matter, so it could be determined that in the Rio Chichico there is low distribution of organic matter, They are very sensitive to chemical contamination, which can denote the incidence of it, for only finding two specimens, many of these families can live in conditions of anoxia, which makes them useful as indicators of high quality, the subject of study in this research.

Family Oligoneuridae 3 specimens, in the sampling we observed this group with very characteristic and hydrodynamic flattened body, which allows them to live in areas of strong current, this group with high thermal requirements, but that can withstand some degree of organic pollution, and found in low numbers agrees with the previous specimens in that in the Rto Chichico the amount of organic matter is low.

Family Perlidae 5 specimens, large, brightly coloured larvae were found, with thoracic gills in plume, this type of specimen lives in streams and rivers of oxygenated water, this family are active predators of other small invertebrates and can become cambales and eat other small ones of their group, in this case being good indicators of environmental quality, it is considered that only for having found 5 specimens, the quality of Rto Chichico is not adequate.

Family Philopotamidae 1 specimen, a single trichoptera was found with free larvae, whose dorsal sclerite and head were orange, in this case they require a high concentration of oxygen so they are excellent indicators of good quality, when only one individual is found in the Rio Chichico it is determined that the environmental quality is low.

Table 2 Results of sampling point 02

KINGDO M	PHYLU M	CLASS	ORDER	FAMILY	ABUNDANCI A		RATE OF EPT PRESENT S
					N	%	
Animali a	Arthropod a	Insect a	Ephemeropter a	Baetidae L.	15	16,48	15
Animali a	Arthropod a	Insect a	Diptera	Ceratopogonide e	4	4,40	0
Animali a	Arthropod a	Insect a	Diptera	Chironomidae	4	4,40	0
Animali a	Arthropod a	Insect a	Neuroptera	Corydalidae	8	8,79	0
Animali a	Arthropod a	Insect a	Coleoptera	Elmidae	9	9,89	0
				Euthyplociidae	1	1,10	1
				Glossosomatide e	2	2,20	2
				Hydrobiosidae	5	5,49	5
Animali a	Arthropod a	Insect a	Trichoptera	Hydropsychida e	4	4,40	4
Animali a	Arthropod a	Insect a	Trichoptera	Leptoceridae	5	5,49	5
Animali a	Arthropod a	Insect a	Trichoptera	Leptohyphidae	5	5,49	5
Animali a	Arthropod a	Insect a	Trichoptera	Leptophlebiida e	1	1,10	1
				Oligoneuridae	2	2,20	2
				Perlidae	1	1,10	1
				Philopotamidae	6	6,59	6
				Psephenidae	5	5,49	0
				Ptilodactylidae	1	1,10	0
				Pyralidae	3	3,30	0
				Other groups	10	10,99	0
				TOTAL	91	100	47
				EPT TOTAL - TOTAL ABUNDANCE	TOTAL ABUNDANCE		■Γ - 91 = 0,51 0,51 x 100 = 51%

Prepared by: The Author

According to the EFA-index methodology, the water quality at sampling point N°02 is determined to be good, as it is rated at 51%.

Graph 4

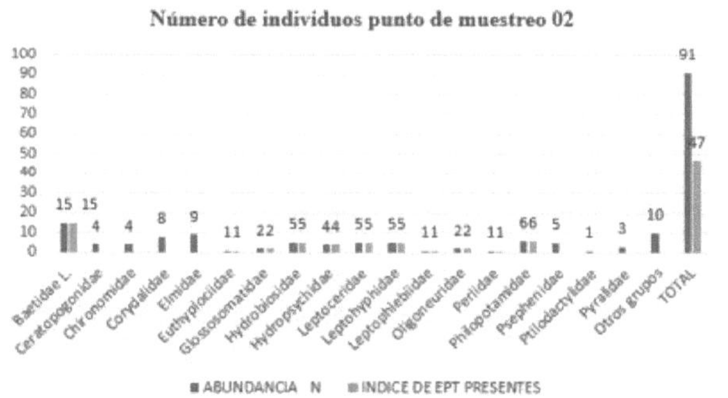

Prepared by: The Author

Interpretation : According to graph 4 similar families were found as in sampling point number one among them: Baetidae 15, Ceratopogonidae 4, Chironomidae 4, Corydalidae 8, Euthyplociidae 1, Glossomatidae 2, Hydropsychidae 5, Leptohyphidae 5, Oligoneuriidae 2, Perlidae 1, Philopotamidae 6 and Leptoceridae 5, for which their analysis has been determined to that developed in the results of sampling point number one, giving importance to the families that were not found in this point as detailed below:

Family: Hydrobiosidae 4 specimens, 5 dark brown cinnamon insects were found in the study site, their average size was 10mm, with characteristics such as: antennae almost as long as their wings, these individuals were found on stones in the river.

Family: Leptophlebiidae 1 specimen. This is the least representative family, with only one specimen found in an area with decomposing leaf litter and fungi; in the case of this individual, being a group that does not tolerate contamination or alterations to riverbank vegetation, they are therefore considered excellent indicators of quality, and it is determined that the water of the Rto Chichico is not of optimum quality.

Family: Psephenidae 5 specimens, these coleoptera were identified by their flattened, oval, pubescent body, with short legs, these specimens being considered as indicators of clean water because of their high oxygen requirements and taking into consideration the low number of specimens, it can be determined that the water quality of the river is not optimal.

Family: Ptilodactylidae 1 specimen, this single specimen was identified by its cylindrical and completely sclerotised body of light brown or relatively dark colour, its long antennae, relatively small thoracic legs, this individual is sensitive to pollution, whose presence in a large population indicates a very good water quality, by having only one specimen it is determined that the water of the Rto is not of good quality.

Family: Pyralidae 3 specimens, these three specimens were found in a stony area, these individuals with good water indicators, their relatively low number denotes the low water quality of the sector.

Table 3 Results of sampling point 03

KINGDOM	PHYLU M	CLASS	ORDER	FAMILY	ABUNDANCE N	%	RATE OF EPT PRESENT S
Animalia	Arthropod a	Insecta	Ephemeropter a	Baetidae	18	38,30	18
Animalia	Arthropod a	Insecta	Diptera	Ceratopogonidae	4	8,51	0
Animalia	Arthropod a	Insecta	Diptera	Chironomidae	2	4,26	0
Animalia	Arthropod a	Insecta	Neuroptera	Corydalidae	6	12,77	0
Animalia	Arthropod a	Insecta	Coleoptera	Elmidae	4	8,51	0
				Hirudinea	4	8,51	0
				Hydrobiosidae	3	6,38	3
				Leptohyphidae	2	4,26	2
				Other groups	4	8,51	0
				TOTAL	47	100	23
				TOTAL EPT - ABUNDANCE TOTAL	TOTAL ABUNDANCE		28- 47 = 0,59 0,59 x 100 = 59%

Prepared by: The Author

According to the EFA index methodology, the water quality at sampling point N°03 is found to be fair, with a value of 59%.

Graph 5

Number of individuals sampling point 03

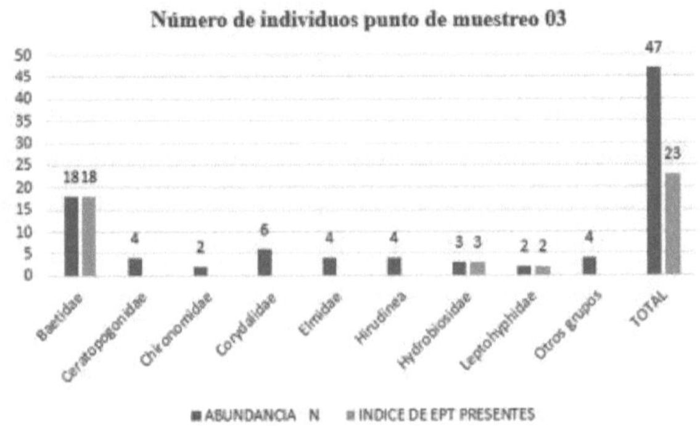

Prepared by: The Author

Interpretation: In graph number five, similar species were identified in sampling points one and two, such as: Baetidae 38.30%, Ceratopogonidae 8.51%, Chironomidae 4.26%, Corydalidae 12.77%, Leptohyphidae 4.26%, where the respective analysis can be found for this case, as new families we have the following:

Family: Hirudinea 4 specimens, this class of phylum anelids were found in low numbers, they were easy to identify as they are the well-known leeches, they are good indicators of environmental quality, as in the case of the other families identified, because of their relatively low population they determine low water quality.

Family: Hydrobiosidae 3 specimens, a very low number of this family were found, the insects found presented characteristics such as: cinnamon to dark brown colour, with areas of light or dark colours in striking patterns, in front wings, these specimens denote water quality, therefore by their number, it is established that the quality is not optimal.

Table 4 Results of sampling point 04

KINGDOM	PHYLUM	CLASS	ORDER	FAMILY	ABUNDANCIA		EPT INDEX PRESENTS
					N	%	
Animalia	Arthropoda	Insecta	Ephemeroptera	Baetidae	10	32,26	10
Animalia	Arthropoda	Insecta	Diptera	Chironomidae	2	6,45	0
Animalia	Arthropoda	Insecta	Neuroptera	Corydalidae	6	19,35	0
Animalia	Arthropoda	Insecta	Coleoptera	Elmidae	4	12,90	0
				Hirudinea	4	12,90	0
Animalia	Arthropoda	Insecta	Trichoptera	Hydrobiosidae	1	3,23	1
Animalia	Arthropoda	Insecta	Trichoptera	Leptohyphidae	2	6,45	2
				Other groups	2	6,45	0
				TOTAL	31	100	13
				EPT TOTAL - TOTAL ABUNDANCE	TOTAL ABUNDANCE		13- 31 = 0,38 0,38 x 100 = 38%

Prepared by: The Author

According to the EFA index methodology, the water quality at sampling point N°04 is determined to be fair, as its value is 38%.

Graph 6

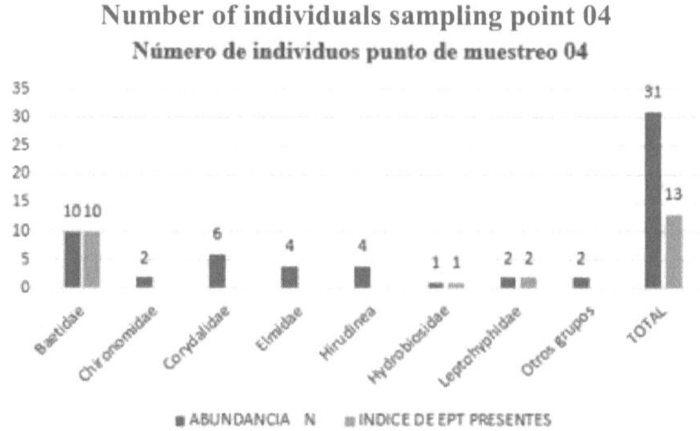

Prepared by: The Author

Interpretation: In graph number six, species similar to the previous points are determined, no new species were distinguished, therefore the analysis is maintained, finding Baetidae 10, Chironomidae 2, Corydalidae 6, Elmidae 4,

Hirudinea 4, Hydrobiosidae 1%, Leptohyphidae 2%, other groups 2.

6.2 To evaluate the physical and chemical quality of the water in the Rio Chichico.

For the present investigation, water samples were taken from the four points indicated above, for which the use of water was established in maximum permissible limits for water for human consumption and domestic use, which only require conventional treatment according to Table 1 of Book VI Annex 1 of the environmental quality and effluent discharge standard of the Unified Text of Secondary Environmental Legislation (see annex photo, 7).

The following permissible limits were determined in the laboratory analysis:

Table 5 Laboratory results at sampling point 01

Determinations	Units	Method	Limit TULAS	Results	COMPLIANCE
pH	Und	4500-B	6 - 9	6,17	YES
Conductivity	LtSiems/cm	2510-B		37,2	-
COD	mg/L	5220-C		1	-
BOD	mg/L	5210-B	2,0	1	YES
Dissolved oxygen	mg/L	4500-O-C	Not less than 6 mg/l	6,2	YES
Nitrates	mg/L	4500-NO3-C	10,0	<0.01	YES
Nitrites	mg/L	4500-NO2-B	1,0	0.009	YES
Phosphates	mg/L	4500-P-D		0.08	-
Iron	mg/L	3500-Fe-D	1,0	0.71	YES
Total dissolved solids	mg/L	2540-C	1.000	14.9	YES

Prepared by: The Author

According to the analysis carried out in the laboratory, it is determined that at sampling point N° 01, all the parameters comply with the permissible limits, but nevertheless there is the presence of organic matter, due to the slight alteration of the biochemical oxygen demand and of the nitrates and nitrites (see annexe photo 11: Point 1).

Organic elements exist in two forms: biodegradable and non-biodegradable. Biodegradable substances are used by micro-organisms as food. In order to oxidise or reduce them, micro-organisms need oxygen. The Biological Oxygen Demand, which is the measure of the amount of oxygen needed by the micro-organisms to degrade the organic elements contained in the water, is therefore detrimental to life itself and to the development of macroinvertebrates.

Presence of nitrates there is a slight presence of nitrates, this parameter affects the balance and decreases the level of oxygen in the water, in this medium nitrates also act as fertilisers for the aquatic vegetation in such a way that, if they concentrate, they can cause the eutrophication of the medium. In a eutrophicated environment there is a proliferation of species such as algae and other green plants that cover the surface. This phenomenon leads to a decrease in the photosynthetic capacity of aquatic organisms, with the consequent death of fish or other aquatic species and loss of biodiversity.

Table 6 Laboratory results at sampling point 02

Determinations	Units	Method	Limit TULAS	Results	COMPLIANCE
pH	Und	4500-B	6 - 9	6,14	YES
Conductivity	LtSiems/cm	2510-B		17	-
COD	mg/L	5220-C		7	-
BOD	mg/L	5210-B	2,0	3	NO
Dissolved oxygen	mg/L	4500-O-C	Not less than 6 mg/l	6,15	YES
Nitrates	mg/L	4500-NO3-C	10,0	<0.01	YES
Nitrites	mg/L	4500-NO2-B	1,0	0.01	YES
Phosphates	mg/L	4500-P-D		0.02	-
Iron	mg/L	3500-Fe-D	1,0	0.64	YES
Total dissolved solids	mg/L	2540-C	1.000	6.8	YES

Prepared by: The Author

According to the analysis carried out in the laboratory, it is determined that at sampling point N° 02, it does not comply with the permissible limits described in the TULAS, in terms of the parameters of biochemical demand for oxygen. (See annex photo 12: -Laboratory analysis Point 2).

This variable shows the anthropogenic influence from the point of view of the affectation by the presence of urban and industrial centres, being the BOD, the amount of oxygen required to degrade the organic matter in the water by biological means, the higher the value the more polluted the water is, therefore, it does not allow the life and development of macroinvertebrates, this may be an explanation for the low number of specimens found.

Table 7 Laboratory results at sampling point 03

Determinations	Units	Method	Limit TULAS	Results	COMPLIANCE
pH	Und	4500-B	6 - 9	6,24	YES
Conductivity	^Siems/c m	2510-B		19.4	-
COD	mg/L	5220-C		19.0	-
BOD	mg/L	5210-B	2,0	16	NO
Dissolved oxygene	mg/L	4500-O-C	Not less than 6 mg/l	5.51	NO
Nitrates	mg/L	4500-NO3-C	10,0	<0.01	YES
Nitrites	mg/L	4500-NO2-B	1,0	0.011	YES
Phosphates	mg/L	4500-P-D		0.01	-
Iron	mg/L	3500-Fe-D	1,0	1.49	NO
Total dissolved solids	mg/L	2540-C	1.000	7.7	YES

Prepared by: The Author

According to the analysis carried out in the laboratory, it is determined that at sampling point N° 03, it does not comply with the biochemical demand for oxygen, dissolved oxygen and iron, according to the TULAS regulations. (See annexe photo 13: Laboratory analysis point 3).

In the same way as at sampling points 1 and 2, this point also does not comply with the permissible BOD limits, and as it is a parameter that allows the determination of the oxygen requirements for the biochemical degradation of the organic matter in the water, its excess does not allow the optimal development of aquatic life.

In the case of Iron, the permissible Hmite is 1mg/l, and the laboratory results give us a value of 1.49 mg/l, the excess of this parameter does not allow the development of life in the water.

Table 8 Laboratory results sampling point 04

Determinations	Units	Method	Limit TULAS	Results	COMPLIANCE
Ph	Und	4500-B	6 - 9	6,04	YES
Conductivity	LtSiems/cm	2510-B		29.3	-
COD	mg/L	5220-C		22.0	-
BOD	mg/L	5210-B	2,0	16.0	NO
Dissolved oxygene	mg/L	4500-O-C	Not less than 6 mg/l	5.83	NO
Nitrates	mg/L	4500-NO3-C	10,0	<0.01	YES
Nitrites	mg/L	4500-NO2-B	1,0	0.011	YES
Phosphates	mg/L	4500-P-D		0.25	-
Iron	mg/L	3500-Fe-D	1,0	1.03	NO
Total dissolved solids	mg/L	2540-C	1.000	11.7	YES

Prepared by: The Author

According to the analysis carried out in the laboratory, it was determined that sampling point number four does not comply with current regulations in terms of biochemical demand for oxygen, dissolved oxygen and iron. (See appendix photo 4: Laboratory analysis of sampling point 4).

4).

In the same way this sampling point also does not meet the BOD, which allows the biological process of natural water purification, therefore the water is polluted with organic matter, decreasing the oxygen necessary for normal aquatic life.

In the case of Dissolved Oxygen, it is the amount of oxygen that is dissolved in the water. It is an indicator of how polluted the water is, i.e. it supports plant and animal life in rivers, at this sampling point it does not meet the permissible limits, at lower permissible limits some fish and macro invertebrates cannot survive.

The presence of iron causes precipitation and undesired colouring, but when it comes into contact with Rtos it persists, and does not condition the proper life of organisms.

6.3 Propose a management plan based on the characterisation of the Rio Chichico riverbed.

6.3.1 Environmental Management Plan

a) Introduction

The Environmental Management Plan (EMP) is a management instrument that allows planning, defining and facilitating the application of environmental measures aimed at preventing, correcting, mitigating and/or compensating the environmental impacts identified in the present investigation, impacts that are generated by the agricultural, livestock and domestic activities carried out daily by the inhabitants living on the banks of the Chichico River in the parish of La Tarqui.

b) Target

To provide the inhabitants of the parish of La Tarqui with environmentally adequate alternatives for the implementation of technically and economically viable environmental management programmes, which should be socialised and applied to the community by the sectional institutions with competence in the environmental area.

c) Outreach

The Environmental Management Plan presents a series of measures to protect, recover and safeguard the water resources of the Chichico River, and at the same time to raise awareness of the importance of the care that should be given to this resource by the inhabitants and authorities of the community of the parish of La Tarqui.

d) Environmental prevention and mitigation programme

The development of the present programme constitutes a basic instrument of environmental management that determines and defines the different actions that allow to prevent, correct, mitigate and/or compensate the environmental impacts generated in the different floristic, faunistic, landscape, recreational and touristic resources, by the action of the different human activities that are carried out daily in the parish of La Tarqui.

- **Objectives**

Propose measures to prevent and mitigate the impacts generated by the different economic and domestic activities of the inhabitants of the parish of La Tarqui.

Table 6 Implement environmental control system for effluents.

Impact N°1: Water contamination of the Chichico River, due to effluent discharges from mechanics, washing machines, car greasing machines.

Type of measure: Prevention and control.

Impact to be controlled: Water contamination of the Rto Chichico. **Objective: To** implement an environmental control system in the mechanics, washing and greasing machines to prevent the contamination of the water of the Rto Chichico.

Mitigation action: The washing machines must comply with the standards regulated in the municipal ordinance for this type of activity, the establishment must have a completely cemented area, under cover, with basins to prevent oil and grease spills, channels for collecting water from the washing process, grease traps and a specific place for the storage of waste, sludge, filters, etc.

Compliance indicators : Number of establishments complying with the environmental standard.	**Means of verification:** Report of inspections by the Environmental Management Unit of the Municipality.
Responsible for implementation, control and monitoring Those responsible for the execution of this measure will be the owner and the Municipal Government of the Canton of Puyo, through the Environmental Management Unit.	
Budget: **Total:** $ 5.000,00.	

Prepared by: The Author.

Box 7 Separate fats and oils from domestic effluents.

Impact N°2: Water contamination of the Chichico River, due to effluent discharge from houses located on the river bank.	
Type of measure: Prevention and control.	
Impact to be controlled: Contamination of water of the Rto Chichico.	
Objective: To separate oils and fats from effluents generated by domestic activities for subsequent discharge into the Chichico River.	
Mitigation action: Dwellings must be provided with a grease trap, oil and grease control device, if necessary construct a waterproof septic tank.	
Compliance indicators : Number of households building a grease trap, or a waterproof septic tank.	**Means of verification:** Report of inspections by the Environmental Management Unit of the Municipality.
Responsible for implementation, control and monitoring Those responsible for the execution of this measure will be the owners of each house and the control and monitoring will be the responsibility of the Municipal Government of the Canton of Puyo, through the Environmental Management Unit.	

Budget: To implement the septic tank and grease trap per house

Total: $ 3.000,00

Prepared by: The Author.

e) Waste Management Programme (WMP)

The aim of waste management is to ensure that solid waste, hazardous waste and solid surpluses are managed with environmental and ecological criteria, so as not to contaminate the Rto Chichico and not to breach current environmental regulations. For this reason, waste needs to be properly managed as indicated in this programme.

- **Objectives**

Generate an environmental management procedure, prioritising the minimisation of waste generation, promoting recycling, reuse and exploitation of waste, and for special and hazardous waste provide effective and safe technical and administrative procedures for its final disposal.

Table 8 Classification of waste and type of containers.

Impact N°3: Pollution by dumping waste in the Chichico River. **Type of measure:** Prevention and control **Objective:** To sort waste and temporarily store it in the suggested containers according to NTE INEN 2841:2014 standard **Action de mitigation:** For temporary storage, waste shall be deposited separately in metal containers, plastic containers or bags and identified with the following colours according to NTE INEN 2841:2014.

TYPE OF WASTE	**CHARACTERISTICS**	**DESCRIPTION**	**CONTAINER TO BE USED**
WASTE ORGANICS	Organic waste susceptible to compostingor biological degradation.	• Waste from food preparation. • Waste from vegetables • Waste from fruits • Waste from crops	- Metal or coloured plastic **GREEN**
WASTE INORGANICS	Non-inert, non-contaminated waste, and susceptible to recycling or re-utilisation.	Polyethylene plastics Cardboard and paper. Glass Wood Ferrous metal	- Metal plastic colour **BLUE**

		Textiles	
WASTE INERTES DANGEROUS	Materials of hazardous use because of their high content of contaminants of chemical origin.	Textiles contaminated with agrochemicals. Used lubricating oils. Agrochemical containers.	- Metal or plastic, colour **BLACK**
WASTE SPECIAL	Waste generated in the medical dispensary, from cures, wounds, etc.	- Gazas - Syringes. • Medicamentos	-
		- Flaskswith waste from medicine.	Metal or coloured plastic **RED**
Compliance indicators: Cantic organic, inorganic, hazardous and special waste		ad de inertes,	**Means of verification:** Recording and reporting of inspections by the Environmental Management Unit of the Municipality.
Responsible for implementation, control and monitoring Those responsible for the execution of this measure will be the owners of each house and the control and monitoring will be the responsibility of the Municipal Government of the Canton of Puyo, through the Environmental Management Unit.			
Budget: To purchase four 0.05 m containers[3] **Total:** $ 280.00/household.			

Prepared by: The Author.

Table 9 Composting of organic fertiliser.

Impact N°4: Contamination by dumping organic waste in the Rto Chichico.

Type of measure: Prevention and control

Objective: To eliminate the negative impact that agriculture has on the Chuchico River, the environment and the health of the workers, through the use of organic fertilisers.

Action de mitigation: Preparation of a basic fermented fertiliser - type "BOCASHI".

Tools:

Shovels, plastic buckets, thermometer, water hose, dust mask and boots.

Time of duration to manufacture it:

Normally, it takes farmers about 20 to 30 days to obtain fermented organic fertilisers at the beginning.

Ingredients:

- 2 quintals of selected common land.
- 2 quintals of rice husks or coffee husks.

• 2 quintals of poultry manure (laying hens). • 1 quintal of broken coal in small pieces. • 10 lbs. of pig or calf manure. • 10 pounds of calcium carbonate or agricultural lime. • 10 pounds of black soil. • 1 book of molasses or purge honey. • 100 grams of baker's yeast, granulated or in loaf form. • Water: according to the puncture test and only once.
Compliance indicators: • Weight of fermented manure. • Production time. • Quantity of raw material available. • Yield of organic fertiliser. | **Means of verification:** Recording and reporting of inspections by the Environmental Management Unit of the Municipality.
Responsible for implementation, control and monitoring The responsible for the execution of this measure will be the owners of each house and the control and monitoring will be carried out by the Municipal Government of the Canton of Puyo, through the Environmental Management Unit.
Budget: To produce fermented manure **Total:** $ 6.00/quintal.

Prepared by: The Author.

Table 10 Sale of materials for recycling.

Impact N°5: Pollution by dumping inorganic waste into the Chichico River.

Type of measure: Prevention and control

Objective: To obtain economic benefit from the reuse or sale of materials for recycling.

Action de mitigation: The following table details the list of waste that should be sorted for reuse or sold for processing or recycling.

Type of waste	Type of waste	United d	Price USD
Paper	Paper	Kg	0,15
	Carton	Kg	0,10

	Newsprint	Kg	0,10
Plastics	PET plastic	Kg	0,20
	HDPE plastic	Kg	0,20
	LDPE plastic	Kg	0,15
	PP Plastic	Kg	0,10
Glass	Glass	Kg	0,25
Non-ferrous scrap	Aluminium	Kg	1,80
	Bronze	Kg	5,00
	Copper	Kg	5.00
Ferrous scrap	Any metal	Kg	2,80

Compliance indicators: • Amount of sorted waste. • Amount of waste sold. • Amount of economic resources generated.	**Means of verification:** Record and report of inspections by the Environmental Management Unit of the Municipality.
Responsible for implementation, control and monitoring The responsible for the execution of this measure will be the owners of each house and the control and monitoring will be carried out by the Municipal Government of the Canton of Puyo, through the Environmental Management Unit.	
Budget: To adapt an area for the sale of recyclable material. **Total:** $ 5000.00/recycler.	

Prepared by: The Author.

f) Training programme

The environmental training programme is a social construction and instrument of transformation where it is inserted; it provides tools for the change of attitudes and behaviours of the inhabitants of the parish of La Tarqui, for a more egalitarian, equitable and democratic society that allows the empowerment of active citizens and promoters of changes that make possible the sustainability of natural resources.

- **Objectives**

Contribute to the organisational strengthening and respond to the need to generate capacities for the adaptation of technical knowledge of environmental management to the inhabitants of the parish of La Tarqui, its officials and authorities in order to contribute to the conservation of the Rto Chichico.

Table 11 Training for the population of the parish of La Tarqui.

Impact N°6: Lack of knowledge of environmental management instruments

Type of measure: Prevention

Objective: To train the population living specifically on the banks of the Rto

Chichico, so that there is a better link between the community and the control authorities, in order to achieve efficient environmental management.

Preventive action: This training will focus on both conservation and enhancement activities for the existing resources on the banks of the Rio Chichicos.

<table>
<tr><th>Content</th><th>Topics</th></tr>
<tr><td>Prohibitions</td><td>Felling of trees on the river bank</td></tr>
<tr><td>Watershed management</td><td>Integrated watershed management.</td></tr>
<tr><td>Environmental Management Plan</td><td>Knowledge of the Environmental Management Plan and its contents.</td></tr>
<tr><td rowspan="4">Environmental Standards</td><td>Environment, concept and definitions</td></tr>
<tr><td>Environmental Protection, EMP and impact minimisation.</td></tr>
<tr><td>Solid and solid waste management and treatment.</td></tr>
<tr><td>Current environmental legislation.</td></tr>
<tr><td>Compliance indicators:
Number of courses and workshops held.</td><td>Means of verification:
Attendance register and report of courses, workshops, etc.</td></tr>
<tr><td colspan="2">Responsible for implementation, control and monitoring
Those responsible for the execution of this measure will be the population and controlled by the Municipal Government of the Canton of Puyo, through the Environmental Management Unit.</td></tr>
<tr><td colspan="2">Budget: For training of 100 people in one day.
Total: $ 2500,00/day.</td></tr>
</table>

Prepared by: The Author.

(g) Rehabilitation plan for affected areas

The measures to be applied in the rehabilitation plan consist of a set of practices to recover the affected areas, facilitating the natural revegetation of the species and the subsequent recovery of the habitat, restoring not only the landscape and vegetation cover but also guaranteeing the stability of the ecosystem.

- **Objectives**

Restore the vegetation of the areas near the river bank in such a way that an ecological balance can be obtained that is compatible with the standards and goals established in the environmental quality norms in accordance with the environmental legal framework in force in the country.

Table 12 Plant production.

Impact N°7: Degraded areas on the banks of the Chichico River

Type of measure: Mitigate

Objective: Production of plants for the revegetation of degraded areas on the banks of the Rio Chichico.

Prevention action:

Production of plants of different species for the revegetation of the intervened areas. The revegetation process will be applied in all the uncovered areas, giving special emphasis to the margin of the Chichico River.

Tools

- Marking tape
- Gloves
- Demolition shovel
- Hoe - pickaxe
- Bars
- Machete
- Pruning shears

The manual sowing methods are:

Triplet planting: A form of planting tree or shrub species that is applied on medium to steep slopes and involves interspersing rows of trees to form a triangle.

Planting in furrows: with a hand tool (hoe), make furrows in the same contour line 40 cm deep, spaced 40 cm apart from the next furrow; the remaining soil is used as a trellis for the furrow. The tree or shrub plant is planted on the trellis.

Live stakes: Stakes obtained from trees or shrubs that are planted by hole planting at distances greater than 50 cm.

Production of nursery plants for revegetation with forestry species

For the propagation and stocking of forest species there will be nurseries, the size of which depends on the capacity of plants to be produced.

Nurseries with an annual production level of 20,000 plants per year are recommended.

Transport of plants from nurseries should be done in boxes (bags of each plant arranged or with soil loaf when produced in plantabanda), not in piles.

Plants should be moved by holding them from the base, it is advisable to grasp them by the stem.

The optimum height of the plants in plastic bags is 30 cm and in plantabanda is 80 cm.

Check that the leaves are healthy and with good foliage, free of pests or diseases.

Compliance indicators: Numberofforestalspecies produced Number of species reforested.	**Means of verification:** Production and reforestation registration and reporting.
Responsible for implementation, control and monitoring Those responsible for the execution of this measure will be the Municipal Government of the Canton of Puyo, through the Environmental Management Unit.	
Budget: For production of forest plants in nurseries to start with a production of 20,000 plants per year of a single species. **Total:** $ 1.200,00/year.(0,60 cts/plantula)	

Prepared by: The Author

h) Environmental Monitoring Programme

An environmental monitoring and follow-up plan shall contain, if applicable, for each phase of the project or activity, the environmental component to be measured and monitored; the associated environmental impact; the location of the monitoring points; the parameters to be used to characterise the state and evolution of the component; the permitted limits; the duration of the monitoring plan for each parameter; the measurement procedure for each parameter; as well as the timing and frequency of submission of the monitoring plan reports to the respective agencies having jurisdiction over the project or activity; the allowable limits; the duration of the monitoring plan for each parameter; the measurement procedure for each parameter; as well as the deadline and frequency of submission of the monitoring plan reports to the respective competent bodies and any other relevant aspects.

Environmental monitoring must be carried out during project activities to ensure that the operations carried out do not have a significant impact on the environment.

- **Target**

To develop an environmental monitoring programme that will allow the periodic, integrated and permanent evaluation of the dynamics of environmental variables, both at the level of the natural environment and the socio-economic and cultural environment, in order to provide accurate and updated information for decision making aimed at the conservation and sustainable use of the Chichico River.

Table 13 Environmental control and monitoring.

Impact N°8: Non-compliance with environmental regulations. **Type of measure:** Control and Mitigate **Objective: To** periodically evaluate each of the environmental programmes to

control the preservation of the Rto Chichico.	
Preventive action: **Environmental prevention and mitigation programme**: Annual water quality monitoring **Waste Management Programme (WMP):** Daily reports of production and sales of recyclable waste and organic fertilisers. **Environmental training programme:** Reports and records of training courses every 2 months. **Programme for rehabilitation of affected areas:** Production report on the number of plants produced each semester. **Environmental Monitoring Programme**: Systematised reporting of all programmes on an annual basis.	
Compliance indicators: Number of programmes monitored.	**Means of verification:** Tracking and monitoring records and reports.
Responsible for implementation, control and monitoring Those responsible for the execution of this measure will be the Municipal Government of the Canton of Puyo, through the Environmental Management Unit.	
Budget: for the implementation of the environmental control and monitoring programme. **Total:** $ 5.000,00/year.	

Prepared by: The Author

Budget breakdown

The following is a summary of all the Programmes contained in this Environmental Management Plan, which are budgeted for an annual period.

Table 9 Budget Breakdown for the Environmental Management Plan

N°	DESCRIPTION	BUDGET REFERENCE
1	Programme for environmental prevention and mitigation	$ 8.000,00
2	Waste Management Programme (WMP)	$ 5.283,00
3	Environmental training programme	$ 15.000,00
4	Programme for rehabilitation of affected areas	$ 500,00
5	Environmental Monitoring Programme	$ 5,000,00
TOTAL		**$ 33.283,00**

Prepared by: The Author

CHAPTER 4

DISCUSSION

7.1 To identify the presence of macroinvertebrates in four points of the Rio Chichico.

Carrera Reyes & Fierro Peralbo (2001), state that macroinvertebrates provide excellent signals about water quality, and by using them in monitoring, the state of the water can be clearly understood: Some of them require good quality water to survive; others, on the other hand, resist, grow and abound when there is pollution. This determines that the sampling carried out in the Chichico river identified 261 species distributed in 5 orders with 20 families, such as: Baetidae, Ceratopogonidae, Chironomidae, Corydalidae , Elmidae, Euthyplociidae, Glossosomatidae, Hydrobiosidae, Hydropsychidae, Hirudinea, Leptoceridae, Leptohyphidae, Leptophlebiidae, Oligoneuridae, Oligochaeta, Perlidae, Philopotamidae, Psephenidae, Ptilodactylidae and Pyralidae. When applying the EPT index (Ephemeroptera, Plecoptera, Trichoptera); according to Carrera Reyes & Fierro Peralbo (2001), these groups are indicators of water quality because they are more sensitive to pollutants: Ephemeroptera or mayflies, Plecoptera or stoneflies and Trichoptera, establishing that as indicators of water quality, in the first two sampling points the water is good, while in the last two sampling points the water is regular, therefore and considering that macroinvertebrates have certain sensitivity to pollutants, the environment in which the families of macroinvertebrates found in the Chichico River develop is not of optimum quality.

7.2 Evaluate the quality of the water resource based on the results found.

The results of the EPT analysis give a certainty that in the lower part of the micro basin of the Rto Chichico, the water is of regular quality, this is due to the numerous discharges of the people to the margin of the river, knowing that in the Unified Text of Secondary Environmental Legislation, Book VI, on water for human and domestic consumption, there are values that do not comply with the maximum permissible limits.

The parameter that most determines the quality of water for the propagation of macroinvertebrates is the Bilogical Oxygen Demand (BOD), the quantity of oxygen that microorganisms, especially bacteria (aerobic or facultative anaerobic, Pseudomonas, Escherichia, Aerobacter, Bacillius), fungi and plankton, consume during the degradation of the organic substances contained in the sample: Pseudomonas, Escherichia, Aerobacter, Bacillius), fungi and plankton, consume during the degradation of the organic substances contained in the sample; According to LLORES (2005) "Water rich in organic matter can cause contamination, and it is worrying that given the valuation according to the

Unified Text of Secondary Environmental Legislation (TULAS), both at sampling points 03 and 04 the BOD is 16 mg/l, with the maximum permissible limits of 2 mg/l, therefore it does not comply with the established in the legislation, therefore being the BOD an indispensable parameter for the quality of the water of rivers, lakes, lagoons or effluents, it is not apt for the survival and the development of the macroinvertebrates, as well as for the human and domestic consumption of the surrounding population. It was determined that the low populations of macroinvertebrates with the EPT index (Ephemeroptera, Plecoptera, Trichoptera), is due to intolerance to anthropogenic disturbance.

7.3 Propose a Management Plan based on the characterisation of the Rio Chichico riverbed.

Given the results obtained in the present investigation, it is necessary to establish corrective measures, in which it contemplates programs that help to improve the environmental conditions and therefore the quality of water of the Rio Chichico; the proposed Environmental Management Plan is oriented to prevent, mitigate and control all the negative impacts generated by anthropogenic activity in the Rio Chichico, after the application of the programs of mitigation, training, surveillance and monitoring, where more important activities are included to the handling of the effluent as to the preservation of macroinvertebrates in the area: mitigation, training, surveillance and monitoring, where more important activities are included to the management of the effluent as to the preservation of macroinvertebrates in the zone; according to the Unified Text of Secondary Legislation of the Environment (TULSMA), in its Book VI, the Environmental Management Plan, allows to establish the actions that are required to prevent, mitigate, control, compensate and correct the possible effects or negative environmental impacts caused in development of a project, work or activity; it also includes the plans of follow-up, evaluation and monitoring and those of contingency.

CHAPTER 5

CONCLUSIONS

- By using the biotic index to measure the water quality of the Chichico river, we obtained data on agricultural and organic pollution, determining the general biological quality of the water, according to the taxonomic composition of the aquatic macroinvertebrates.
- By means of the EPT index (Ephemeroptera, Plecoptera, Trichoptera), it helped to determine that the Chichico River has a greater impact of contamination that affects the communities of the families of these orders, therefore the quality of the water is regular, this is due to the numerous discharges of the houses that are on the banks of the river.
- The physical and chemical parameters analysed in the Chichico river indicate that the quality of the water is regular, this coincides with the result of the EPT Index, a result that is assumed due to the high load of organic matter; water that is not suitable for the survival and development of macroinvertebrates, as well as for human and domestic consumption of the surrounding population.
- The physico-chemical and biological results obtained in the present study are very useful for the evaluation of the water quality of the Chichico river, especially for the use of aquatic bioindicators as a tool of easy application, low cost and with efficient results.
- The study serves for the creation of a fundamental database to know the composition of the communities of the aquatic biodiversity of macroinvertebrates in the Chichico river in the parish of La Tarqui.
- There are no similar studies, they lack an Environmental Management Plan, a plan that, in a detailed manner, establishes programmes to prevent, mitigate, control, compensate and correct the possible environmental effects or impacts caused by the different economic and domestic activities produced by the inhabitants living on the banks of the Chichico River.

RECOMMENDATIONS

- To promote the use of this methodology in the basins, sub-basins and micro-basins of our region, in order to enrich the list of families and genera of the region and thus better know the tolerance and sensitivity values of these to different levels of contamination.
- To integrate the use of aquatic macroinvertebrates in water quality studies into the Environmental Management and Conservation Engineering Degree, with qualified technical personnel for taxonomic identification and thus contribute to this growing and popular technology in the environmental area.
- Propose a programme for the management of organic and inorganic waste, in order to control the contamination of the Chichico River.
- The construction of devices for the separation of fats and oils of domestic origin and of service establishments, in this way recovering and preserving the water resource and at the same time being used by its inhabitants for agriculture, tourism and human consumption.
- Propose an Environmental Management Plan, a management instrument for the adequate management of natural resources in the Rto Chichico, in the parish of La Tarqui in the city of Puyo, which consists of eight programmes that must be socialised and executed jointly with the inhabitants and authorities of the Municipality of Puyo.

BIBLIOGRAPHY

(FAO)2005. (n.d.). POLAR SCIENCE.COM. Retrieved from http://www.cienciapopular.com/ecologia/deforestacion-mundial

SMITH, R. L., & SMITH , T. (2007). Ecologia Sexta Edicion (Madrid ed.). Sixth Edition: ADDISON-WESLEY.

(FAO)2005. (n.d.). POLAR SCIENCE.COM. Retrieved from http://www.cienciapopular.com/ecologia/deforestacion-mundial

(2005). In A. V. ESPINOZA G, Fundamentos de Evaluacion de Impactos Ambientales. Santiago - Chile.

2005, O. d. (1 January 2007). POPULARSCIENCE.COM. Retrieved 12 MARCH 2014, from http://www.cienciapopular.com/ecologia/deforestacion-mundial

2010, M. (2010). Aprovechamiento de los Recursos Forestales 20072009. Quito-Ecuador.

BARBA, L. E. (2005). BASIC CONCEPTS OF WATER POLLUTION AND MEASUREMENT PARAMETERS. Retrieved from www.bvsde.paho.org/bvsaar/e/fulltext/gestion/conceptos.pdf

BARBER, H. e. (2008). Global Diversity of Mayflies (Ephemeroptera,Insecta) in Freshwater. Retrieved from http://link.springer.com/content/pdf/10.1007%2Fs10750-007- 9028-y

Biosfera Gestion Ambiental (12 2009). Estudio de Impacto Ambiental Ex-Post Centro de acopio de cilindros de GLP- Esmeraldas. Retrieved 02 07 07 2014, from http://www.biosfera.com.ec/fckeditor_upload/File/BIOSFERA/PETROCOMERCIAL/CAP_VIII.pdf

Boxshall, G. A. (2006). The Plunder of Marine Biodiversity, Scientific and Technological Challenges.

Bromley, W. y. (2005). Methodology of Environmental Impact Assessment.

Bustos F. (2010). Manual de Gestion y Control Ambiental (Third ed.). Ecuador: R.N Industria Grafica.

carrera reyes, c., & fierro peralbo, k. (2005). Macroinvertebrates as indicators of water quality. Quito.

Clara, M. R. (2005). Analysis of the quality of water for human consumption and local perception of appropriate technologies for disinfection at household scale in the El Limon micro-watershed, San Jeronimo, Honduras.

clau. (n.d.). contaminates dela gua . Retrieved from http://html.contaminacion-del-agua_11.html CLIRSEN . (2005).

CORMA (2005). 31.

DAJOZ, R. (2005). TRATADO DE ECOLOGIA (2ª ED.) (Vol. Segundo). MUNDI-PRENSA LIBROS, S.A.

Dommguez, E. &. (2009). Benthic Macroinvertebrates
South Americans. Systematics and Biology. Argentina.
ECOLAP and MAE. (2007). Grna del Patrimonio de Areas Naturales Protegidas del Ecuador. Quito: IGM.
FAO. (2005). Retrieved from
https://www.oas.org/DSD/publications/Unit/oea19s/ch008.htm
FAO. (2005). Retrieved from
https://www.oas.org/DSD/publications/Unit/oea19s/ch009.htm#
3.2%20wood%20sawed%20wood
FERNANDEZ-VITORA, V. C. (2005). "GUIA METODOLOGICA PARA LA EVALUACION DEL IMPACTO AMBIENTAL" (second edition ed.). Madrid-Espana: MUNDI-PRENSA.
Figueroa, R. V. (2005). Benthic macroinvertebrates as indicators of chilean quality. Revista Chilena de Historia Natural, 275-285.
GORDILLO, O. (07 December 2009). Ecolog^a del Ecuador. Retrieved 26 October 2011, from
http://ogordillo.blogspot.com/2008/09/ecologa-para- students-of-tourism.html
GRIJALVA, T., & OTALVARO, J. (2010). Zonificacion Ecologica - Ambiental del Canton Pimampiro. Ibarra.
GUARANDA, W. (16 November 2009). Fundacion Regional de Asesoria en Derechos Humanos. Retrieved 26 October 2011, from Auntessobrela actividad petrolera:
http://www.inredh.org/index.php?option=com_content&view=a
rticle&id=288%3Aexplotacion-petrolera-en-el- ecuador&Itemid=126
Guerrero, I. L. (2005). ENVIRONMENTAL IMPACT ASSESSMENT AND PROPOSAL OF THE.
Gutierrez, P. (2010). Plecoptera. Retrieved from Revista de Biolog^a Tropical:
http://www.scielo.sa.cr/scielo.php?script=sci_arttext&pid=S003 4-
77442010000800006&lng=en&nrm=iso>. ISSN 0034-7744.
Holzenthal, R. B. (2007). Order Trichoptera Kirby. Retrieved from Linnaeus Tercentenary: Progress in Invertebrate:
http://www.revistas.ucr.ac.cr/index.php/rbt/article/view/5413/51 59
Irene Campos Gomez, U. E. (n.d.). Environmental Sanitation. EUNED.
leopold (2005).
book 6. (2008). Quito.
Lloret, P. (2005). Training Systems for the Management of Natural Resources. Retrieved January 20, 2014, from GestiondeCuencas Hidrograficas:
http://www.avina.net/esp/wp- content/uploads/2013/03/MODULO-5-OK.pdf
Ministry of Environment, V. y. (2006). Rural Plumbing. Retrieved January 21,

2014, from Programa Cultura Empresarial 9: http://www.avina.net/esp/wp-content/uploads/2013/03/MODULO-5-OK.pdf
Federal Ministry for Economic Cooperation and Development (BMZ) (2005). Guide to Environmental Protection (Vol. Volume II). Germany.
MINISTRY, D. A. (2005). Bosque Protector "Tsuraku o Arutam". Quito: MAE.
Murillo, O., villanueva, A., r, h., Heredia, S., Ojeda, A., Martinez, S., & Alcantara, I. (2009). The periphyton of a hypereutrophic hyposaline lake in Michoacan, Mexico.
NAREDO, J., & PARRA, F. (2005). Towards a science of natural resources. Madrid: Siglo XXI.
Nogales, S. R. (2005). ENVIRONMENTAL IMPACT ASSESSMENT AND ENVIRONMENTAL MANAGEMENT PLAN PROPOSAL FOR THE AMBUQU IRRIGATION PROJECT.
WHO. (2005). Guidelines for Drinking Water Quality. Retrieved 05 February 2014
WHO. (2005). Manual on the prevention of waterborne diseases. Retrieved February 04, 2014, from file:///C:/Users/USERS/USER/Downloads/Calidad%20de%20de%20Servi%20del%20Agua%20Potable%20en%20la%20Ciudad%20d e%20Portovi%20.pdf
ORBE, A. P. (2009). Analysis of the inhibition of reactive clays and unstable shales of the Orteguanza, Tiyuyacu, Tena and Napo (Shale) formations with amine-based fluids in a field in eastern Ecuador. Quito: Universidad Poli'tecnica Nacional.
Othon, Z. M. (2005). La calidad y Servicios de Agua Potable en Portoviejo. Retrieved January 27, 2014, from www.ozpatra@ecua.net.ec
PACHECO, J. X. (2010). Fallas Geologicas en el Ecuador. Loja: Universidad Tecnica Particular de Loja.
Perez, A., & Rodriguez, A. (2008). Physicochemical index of water quality for the management of tropical flood ponds. Revista Biologica Tropical, 56.
Perez, L. B. (2004). Water resources management.
Porras (2005). Arte Rupestre Alto Napo-Valle del Misahualli,. Quito, Ecuador : Artes Graficas Senal Impresenal Cia. Ltda.
prezi (2005). Retrieved April 28, 2014, from http://prezi.com/aruhmiq2c4gb/biota-en-recursos-hidricos/
Restrepo Claudia, P. (2008). Water, health and life. Ministry of Environment, Housing. Retrieved February 02, 2014, from ViceministeriodeAguaySaneamiento : http://www.avina.net/esp/wp- content/uploads/2013/03/MODULO-5-OK.pdf
RIVADENEIRA, B. (2005). The Oriente Basin: Tectonic style, stages of

deformation and geological characteristics of the main petroproduction fields . Quito: Escuela Politecnica Nacional.
Roldan, G. &. (2008). Fundamentals of Neotropical Limnology (2nd Edition).
SENPLADES - CAF. (2005). Plan Estrategico para la Reduccion del Riesgo en el Territorio Ecuatoriano. Quito: SENPLADES - CAF.
SIERRA, R. (2005). Propuesta Preliminar de un sistema de Clasificacion de Vegetacion para el Ecuador Continental. Quito: INEFAN/GEF-BIRF and EcoCiencia Project.
Springer, M. (2010). Trichoptera. Journal of Tropical Biology, 150152.
TULAS (2010). Book VI, Environmental Quality Standard and Effluent Discharge: Water Resource.
UNESCO, F. (2005). World Soil Gasification. FAO.
V . CONESA FDEZ - VTI'ORA. (2010). Gina Metodologica para la Evaluación del Impacto Ambiental (4ª - 2010 ed.). Madrid- Spain: Grupo Mundil-Prensa.
VI G.G, a. (11 March 2003). MIMOSA.PCNTICMES.ES. Retrieved from http://mimosa.pntic.mec.es/vgarci14/index.htm
VII T. L. (2010). Environmental Quality Standard and Effluent Discharge: Water Resource.
WINCKELL, A., & ZEBROESKI, C. (2005). The Natural Landscapes of Ecuador. The Regions and Landscape of Ecuador. Quito: IPGH .
WINCKELL, A., & ZEBROWSKI, C. (2005). The Natural Landscapes of Ecuador. The regions and landscapes of Ecuador. Quito: IPGH.
ZAPATA, R. (2006). Universidad Nacional de Colombia- Sede Medellrn. Retrieved 27 October 2011, from Qrnmica de los procesos pedogeneticos:

ANNEXES

Table 10

UNIVERSIDAD NACIONAL DE LOJA
PLAN DE CONTINGENCIA
CARRERA DE INGENIERÍA EN MANEJO Y CONSERVACIÓN DEL MEDIO AMBIENTE

Hoja de campo:
Sitio de colección:
Nombre del río:
Fecha de colección:
Técnico a cargo:

Coordenadas:

x	y

CLASIFICACIÓN	ABUNDANCIAS (NÚMERO DE INDIVIDUOS)	EPT PRESENTES
Anisoptera		
Bivalvia		
Baetidae		
Ceratopogonidae		
Chironomidae		
Corydalidae		
Elmidae		
Euthyplociidae		
Gastropoda		
Glossosomatidae		
Gordioidea		
Hirudinea		
Hydrachnidae		
Hydrobiosidae		
Hydropsychidae		
Leptoceridae		
Leptohyphidae		
Leptophlebiidae		
Naucoridae		
Oligochaeta		
Oligoneuridae		
Perlidae		
Philopotamidae		
Psephenidae		
Ptilodactylidae		
Pyralidae		
Simuliidae		
Tipulidae		
Turbellaria		
Veliidae		
Zygoptera		
Otros grupos		
TOTAL:		

CALIDAD DE AGUA
75 - 100% Muy buena
50 - 74% Buena
25 - 49% Regular
0 - 24% Mala

Mice ETP

PHOTOGRAPHIC ANNEXES

Photo 2

First sampling point

Photo 3

Second sampling point

Photo 4

Third sampling point
Photo 5

Fourth sampling point
Photo 6

Water sample collection
Photo 7

Water sampling
Photo 8

Sample delivery to the laboratory (ESPOCH)
Photo 9

Collection of macroinvertebrates
Photo 10

Macroinvertebrate collection with the Surber net

Collection of macroinvertebrates

Photo 10

[illegible]

MIX
Papier aus verantwortungsvollen Quellen
Paper from responsible sources
FSC® C105338

Printed by Books on Demand GmbH, Norderstedt / Germany